人工智能：Python实现（影印版）
Artificial Intelligence with Python

Prateek Joshi 著

南京　东南大学出版社

图书在版编目(CIP)数据

人工智能:Python 实现:英文/(美)普拉提克·乔希(Prateek Joshi)著. —影印本. —南京:东南大学出版社,2017.10(2024.1重印)

书名原文:Artificial Intelligence with Python

ISBN 978-7-5641-7358-6

Ⅰ.①人… Ⅱ.①普… Ⅲ.①人工智能-英文 Ⅳ.①TP18

中国版本图书馆 CIP 数据核字(2017)第 193636 号

图字:10-2017-121 号

© 2017 by PACKT Publishing Ltd

Reprint of the English Edition, jointly published by PACKT Publishing Ltd and Southeast University Press, 2017. Authorized reprint of the original English edition, 2017 PACKT Publishing Ltd, the owner of all rights to publish and sell the same.

All rights reserved including the rights of reproduction in whole or in part in any form.

英文原版由 PACKT Publishing Ltd 出版 2017。

英文影印版由东南大学出版社出版 2017。此影印版的出版和销售得到出版权和销售权的所有者——PACKT Publishing Ltd 的许可。

版权所有,未得书面许可,本书的任何部分和全部不得以任何形式重制。

人工智能:Python 实现(影印版)

出版发行:东南大学出版社
地　　址:南京四牌楼 2 号　　邮编:210096
出 版 人:江建中
网　　址:http://www.seupress.com
电子邮件:press@seupress.com
印　　刷:苏州市古得堡数码印刷有限公司
开　　本:787 毫米×980 毫米　16 开本
印　　张:28
字　　数:548 千字
版 印 次:2024年1月第1版第5次印刷
书　　号:ISBN 978-7-5641-7358-6
定　　价:108.00元

本社图书若有印装质量问题,请直接与营销部联系。电话(传真):025-83791830

Credits

Author

Prateek Joshi

Reviewer

Richard Marsden

Commissioning Editor

Veena Pagare

Acquisition Editor

Tushar Gupta

Content Development Editor

Aishwarya Pandere

Technical Editor

Karan Thakkar

Copy Editors

Vikrant Phadkay

Safis Editing

Project Coordinator

Nidhi Joshi

Proofreader

Safis Editing

Indexer

Mariammal Chettiyar

Production Coordinator

Shantanu N. Zagade

About the Author

Prateek Joshi is an artificial intelligence researcher, published author of five books, and TEDx speaker. He is the founder of Pluto AI, a venture-funded Silicon Valley startup building an analytics platform for smart water management powered by deep learning. His work in this field has led to patents, tech demos, and research papers at major IEEE conferences. He has been an invited speaker at technology and entrepreneurship conferences including TEDx, AT&T Foundry, Silicon Valley Deep Learning, and Open Silicon Valley. Prateek has also been featured as a guest author in prominent tech magazines.

His tech blog (www.prateekjoshi.com) has received more than 1.2 million page views from 200 over countries and has over 6,600+ followers. He frequently writes on topics such as artificial intelligence, Python programming, and abstract mathematics. He is an avid coder and has won many hackathons utilizing a wide variety of technologies. He graduated from University of Southern California with a master's degree specializing in artificial intelligence. He has worked at companies such as Nvidia and Microsoft Research. You can learn more about him on his personal website at www.prateekj.com.

About the Reviewer

Richard Marsden has over 20 years of professional software development experience. After starting in the field of geophysical surveying for the oil industry, he has spent the last ten years running the Winwaed Software Technology LLC independent software vendor. Winwaed specializes in geospatial tools and applications including web applications, and operates the `http://www.mapping-tools.com` website for tools and add-ins for geospatial applications such as Caliper Maptitude and Microsoft MapPoint.

Richard was also a technical reviewer of the following Packt publications: *Python Geospatial Development* and *Python Geospatial Analysis Essentials*, both by *Erik Westra*; *Python Geospatial Analysis Cookbook* by *Michael Diener*; *Mastering Python Forensics* by *Drs Michael Spreitzenbarth* and *Dr Johann Uhrmann*; and *Effective Python Penetration Testing* by *Rejah Rehim*.

www.PacktPub.com

For support files and downloads related to your book, please visit `www.PacktPub.com`.

Did you know that Packt offers eBook versions of every book published, with PDF and ePub files available? You can upgrade to the eBook version at `www.PacktPub.com` and as a print book customer, you are entitled to a discount on the eBook copy. Get in touch with us at `service@packtpub.com` for more details.

At `www.PacktPub.com`, you can also read a collection of free technical articles, sign up for a range of free newsletters and receive exclusive discounts and offers on Packt books and eBooks.

`https://www.packtpub.com/mapt`

Get the most in-demand software skills with Mapt. Mapt gives you full access to all Packt books and video courses, as well as industry-leading tools to help you plan your personal development and advance your career.

Why subscribe?

- Fully searchable across every book published by Packt
- Copy and paste, print, and bookmark content
- On demand and accessible via a web browser

Customer Feedback

Thank you for purchasing this Packt book. We take our commitment to improving our content and products to meet your needs seriously—that's why your feedback is so valuable. Whatever your feelings about your purchase, please consider leaving a review on this book's Amazon page. Not only will this help us, more importantly it will also help others in the community to make an informed decision about the resources that they invest in to learn.

You can also review for us on a regular basis by joining our reviewers' club. **If you're interested in joining, or would like to learn more about the benefits we offer, please contact us**: customerreviews@packtpub.com.

Table of Contents

Preface	1
Chapter 1: Introduction to Artificial Intelligence	7
What is Artificial Intelligence?	8
Why do we need to study AI?	8
Applications of AI	12
Branches of AI	14
Defining intelligence using Turing Test	16
Making machines think like humans	18
Building rational agents	20
General Problem Solver	21
Solving a problem with GPS	22
Building an intelligent agent	22
Types of models	24
Installing Python 3	24
Installing on Ubuntu	25
Installing on Mac OS X	25
Installing on Windows	26
Installing packages	26
Loading data	27
Summary	29
Chapter 2: Classification and Regression Using Supervised Learning	31
Supervised versus unsupervised learning	31
What is classification?	32
Preprocessing data	33
Binarization	33
Mean removal	34
Scaling	35
Normalization	36
Label encoding	37
Logistic Regression classifier	38
Naïve Bayes classifier	43
Confusion matrix	47
Support Vector Machines	50
Classifying income data using Support Vector Machines	52

What is Regression?	55
Building a single variable regressor	56
Building a multivariable regressor	59
Estimating housing prices using a Support Vector Regressor	61
Summary	63

Chapter 3: Predictive Analytics with Ensemble Learning — 65

What is Ensemble Learning?	65
Building learning models with Ensemble Learning	66
What are Decision Trees?	66
Building a Decision Tree classifier	67
What are Random Forests and Extremely Random Forests?	72
Building Random Forest and Extremely Random Forest classifiers	72
Estimating the confidence measure of the predictions	78
Dealing with class imbalance	82
Finding optimal training parameters using grid search	89
Computing relative feature importance	92
Predicting traffic using Extremely Random Forest regressor	95
Summary	98

Chapter 4: Detecting Patterns with Unsupervised Learning — 99

What is unsupervised learning?	99
Clustering data with K-Means algorithm	100
Estimating the number of clusters with Mean Shift algorithm	106
Estimating the quality of clustering with silhouette scores	109
What are Gaussian Mixture Models?	114
Building a classifier based on Gaussian Mixture Models	115
Finding subgroups in stock market using Affinity Propagation model	120
Segmenting the market based on shopping patterns	122
Summary	126

Chapter 5: Building Recommender Systems — 127

Creating a training pipeline	127
Extracting the nearest neighbors	130
Building a K-Nearest Neighbors classifier	134
Computing similarity scores	141
Finding similar users using collaborative filtering	145
Building a movie recommendation system	148
Summary	151

Chapter 6: Logic Programming — 153

What is logic programming?	153
Understanding the building blocks of logic programming	156
Solving problems using logic programming	156
Installing Python packages	157
Matching mathematical expressions	157
Validating primes	159
Parsing a family tree	161
Analyzing geography	167
Building a puzzle solver	170
Summary	174

Chapter 7: Heuristic Search Techniques — 175

What is heuristic search?	175
Uninformed versus Informed search	176
Constraint Satisfaction Problems	177
Local search techniques	177
Simulated Annealing	178
Constructing a string using greedy search	179
Solving a problem with constraints	183
Solving the region-coloring problem	186
Building an 8-puzzle solver	189
Building a maze solver	194
Summary	199

Chapter 8: Genetic Algorithms — 201

Understanding evolutionary and genetic algorithms	201
Fundamental concepts in genetic algorithms	202
Generating a bit pattern with predefined parameters	203
Visualizing the evolution	210
Solving the symbol regression problem	219
Building an intelligent robot controller	224
Summary	231

Chapter 9: Building Games With Artificial Intelligence — 233

Using search algorithms in games	234
Combinatorial search	234
Minimax algorithm	235
Alpha-Beta pruning	235
Negamax algorithm	236
Installing easyAI library	236
Building a bot to play Last Coin Standing	237

Building a bot to play Tic-Tac-Toe	241
Building two bots to play Connect Four™ against each other	244
Building two bots to play Hexapawn against each other	248
Summary	252

Chapter 10: Natural Language Processing — 253

Introduction and installation of packages	253
Tokenizing text data	255
Converting words to their base forms using stemming	256
Converting words to their base forms using lemmatization	258
Dividing text data into chunks	260
Extracting the frequency of terms using a Bag of Words model	262
Building a category predictor	265
Constructing a gender identifier	268
Building a sentiment analyzer	271
Topic modeling using Latent Dirichlet Allocation	275
Summary	278

Chapter 11: Probabilistic Reasoning for Sequential Data — 279

Understanding sequential data	279
Handling time-series data with Pandas	280
Slicing time-series data	283
Operating on time-series data	285
Extracting statistics from time-series data	288
Generating data using Hidden Markov Models	292
Identifying alphabet sequences with Conditional Random Fields	295
Stock market analysis	300
Summary	303

Chapter 12: Building A Speech Recognizer — 305

Working with speech signals	305
Visualizing audio signals	306
Transforming audio signals to the frequency domain	309
Generating audio signals	311
Synthesizing tones to generate music	314
Extracting speech features	316
Recognizing spoken words	320
Summary	326

Chapter 13: Object Detection and Tracking — 327

| Installing OpenCV | 328 |

Frame differencing	328
Tracking objects using colorspaces	331
Object tracking using background subtraction	335
Building an interactive object tracker using the CAMShift algorithm	339
Optical flow based tracking	347
Face detection and tracking	354
Using Haar cascades for object detection	354
Using integral images for feature extraction	355
Eye detection and tracking	358
Summary	361

Chapter 14: Artificial Neural Networks — 363

Introduction to artificial neural networks	363
Building a neural network	364
Training a neural network	364
Building a Perceptron based classifier	365
Constructing a single layer neural network	369
Constructing a multilayer neural network	373
Building a vector quantizer	378
Analyzing sequential data using recurrent neural networks	381
Visualizing characters in an Optical Character Recognition database	385
Building an Optical Character Recognition engine	388
Summary	391

Chapter 15: Reinforcement Learning — 393

Understanding the premise	393
Reinforcement learning versus supervised learning	394
Real world examples of reinforcement learning	395
Building blocks of reinforcement learning	396
Creating an environment	397
Building a learning agent	402
Summary	406

Chapter 16: Deep Learning with Convolutional Neural Networks — 407

What are Convolutional Neural Networks?	407
Architecture of CNNs	408
Types of layers in a CNN	409
Building a perceptron-based linear regressor	410
Building an image classifier using a single layer neural network	416
Building an image classifier using a Convolutional Neural Network	418
Summary	424

Index

Preface

Artificial intelligence is becoming increasingly relevant in the modern world where everything is driven by data and automation. It is used extensively across many fields such as image recognition, robotics, search engines, and self-driving cars. In this book, we will explore various real-world scenarios. We will understand what algorithms to use in a given context and write functional code using this exciting book.

We will start by talking about various realms of artificial intelligence. We'll then move on to discuss more complex algorithms, such as Extremely Random Forests, Hidden Markov Models, Genetic Algorithms, Artificial Neural Networks, and Convolutional Neural Networks, and so on. This book is for Python programmers looking to use artificial intelligence algorithms to create real-world applications. This book is friendly to Python beginners, but familiarity with Python programming would certainly be helpful so you can play around with the code. It is also useful to experienced Python programmers who are looking to implement artificial intelligence techniques.

You will learn how to make informed decisions about the type of algorithms you need to use and how to implement those algorithms to get the best possible results. If you want to build versatile applications that can make sense of images, text, speech, or some other form of data, this book on artificial intelligence will definitely come to your rescue!

What this book covers

Chapter 1, *Introduction to Artificial Intelligence*, teaches you various introductory concepts in artificial intelligence. It talks about applications, branches, and modeling of Artificial Intelligence. It walks the reader through the installation of necessary Python packages.

Chapter 2, *Classification and Regression Using Supervised Learning*, covers various supervised learning techniques for classification and regression. You will learn how to analyze income data and predict housing prices.

Chapter 3, *Predictive Analytics with Ensemble Learning*, explains predictive modeling techniques using Ensemble Learning, particularly focused on Random Forests. We will learn how to apply these techniques to predict traffic on the roads near sports stadiums.

Chapter 4, *Detecting Patterns with Unsupervised Learning*, covers unsupervised learning algorithms including K-means and Mean Shift Clustering. We will learn how to apply these algorithms to stock market data and customer segmentation.

Preface

Chapter 5, *Building Recommender Systems*, illustrates algorithms used to build recommendation engines. You will learn how to apply these algorithms to collaborative filtering and movie recommendations.

Chapter 6, *Logic Programming*, covers the building blocks of logic programming. We will see various applications, including expression matching, parsing family trees, and solving puzzles.

Chapter 7, *Heuristic Search Techniques*, shows heuristic search techniques that are used to search the solution space. We will learn about various applications such as simulated annealing, region coloring, and maze solving.

Chapter 8, *Genetic Algorithms*, covers evolutionary algorithms and genetic programming. We will learn about various concepts such as crossover, mutation, and fitness functions. We will then use these concepts to solve the symbol regression problem and build an intelligent robot controller.

Chapter 9, *Building Games with Artificial Intelligence*, teaches you how to build games with artificial intelligence. We will learn how to build various games including Tic Tac Toe, Connect Four, and Hexapawn.

Chapter 10, *Natural Language Processing*, covers techniques used to analyze text data including tokenization, stemming, bag of words, and so on. We will learn how to use these techniques to do sentiment analysis and topic modeling.

Chapter 11, *Probabilistic Reasoning for Sequential Data*, shows you techniques used to analyze time series and sequential data including Hidden Markov models and Conditional Random Fields. We will learn how to apply these techniques to text sequence analysis and stock market predictions.

Chapter 12, *Building A Speech Recognizer*, demonstrates algorithms used to analyze speech data. We will learn how to build speech recognition systems.

Chapter 13, *Object Detection and Tracking*, It covers algorithms related to object detection and tracking in live video. We will learn about various techniques including optical flow, face tracking, and eye tracking.

Chapter 14, *Artificial Neural Networks*, covers algorithms used to build neural networks. We will learn how to build an Optical Character Recognition system using neural networks.

Chapter 15, *Reinforcement Learning*, teaches the techniques used to build reinforcement learning systems. We will learn how to build learning agents that can learn from interacting with the environment.

Chapter 16, *Deep Learning with Convolutional Neural Networks*, covers algorithms used to build deep learning systems using Convolutional Neural Networks. We will learn how to use TensorFlow to build neural networks. We will then use it to build an image classifier using convolutional neural networks.

What you need for this book

This book is focused on artificial intelligence in Python as opposed to the Python itself. We have used Python 3 to build various applications. We focus on how to utilize various Python libraries in the best possible way to build real world applications. In that spirit, we have tried to keep all of the code as friendly and readable as possible. We feel that this will enable our readers to easily understand the code and readily use it in different scenarios.

Who this book is for

This book is for Python developers who want to build real-world artificial intelligence applications. This book is friendly to Python beginners, but being familiar with Python would be useful to play around with the code. It will also be useful for experienced Python programmers who are looking to use artificial intelligence techniques in their existing technology stacks.

Conventions

In this book, you will find a number of text styles that distinguish between different kinds of information. Here are some examples of these styles and an explanation of their meaning.

Code words in text, database table names, folder names, filenames, file extensions, pathnames, dummy URLs, user input, and Twitter handles are shown as follows: "We can include other contexts through the use of the `include` directive."

A block of code is set as follows:

```
[default]
exten => s,1,Dial(Zap/1|30)
exten => s,2,Voicemail(u100)
exten => s,102,Voicemail(b100)
exten => i,1,Voicemail(s0)
```

When we wish to draw your attention to a particular part of a code block, the relevant lines or items are set in bold:

```
[default]
exten => s,1,Dial(Zap/1|30)
exten => s,2,Voicemail(u100)
exten => s,102,Voicemail(b100)
exten => i,1,Voicemail(s0)
```

Any command-line input or output is written as follows:

```
# cp /usr/src/asterisk-addons/configs/cdr_mysql.conf.sample
      /etc/asterisk/cdr_mysql.conf
```

New terms and **important words** are shown in bold. Words that you see on the screen, for example, in menus or dialog boxes, appear in the text like this: "The shortcuts in this book are based on the `Mac OS X 10.5+` scheme."

Warnings or important notes appear in a box like this.

Tips and tricks appear like this.

Reader feedback

Feedback from our readers is always welcome. Let us know what you think about this book-what you liked or disliked. Reader feedback is important for us as it helps us develop titles that you will really get the most out of.

To send us general feedback, simply e-mail `feedback@packtpub.com`, and mention the book's title in the subject of your message.

If there is a topic that you have expertise in and you are interested in either writing or contributing to a book, see our author guide at `www.packtpub.com/authors`.

Customer support

Now that you are the proud owner of a Packt book, we have a number of things to help you to get the most from your purchase.

Downloading the example code

You can download the example code files for this book from your account at http://www.packtpub.com. If you purchased this book elsewhere, you can visit http://www.packtpub.com/support and register to have the files e-mailed directly to you.

You can download the code files by following these steps:

1. Log in or register to our website using your e-mail address and password.
2. Hover the mouse pointer on the **SUPPORT** tab at the top.
3. Click on **Code Downloads & Errata**.
4. Enter the name of the book in the **Search** box.
5. Select the book for which you're looking to download the code files.
6. Choose from the drop-down menu where you purchased this book from.
7. Click on **Code Download**.

Once the file is downloaded, please make sure that you unzip or extract the folder using the latest version of:

- WinRAR / 7-Zip for Windows
- Zipeg / iZip / UnRarX for Mac
- 7-Zip / PeaZip for Linux

The code bundle for the book is also hosted on GitHub at https://github.com/PacktPublishing/Artificial-Intelligence-with-Python. We also have other code bundles from our rich catalog of books and videos available at https://github.com/PacktPublishing/. Check them out!

Downloading the color images of this book

We also provide you with a PDF file that has color images of the screenshots/diagrams used in this book. The color images will help you better understand the changes in the output. You can download this file from https://www.packtpub.com/sites/default/files/downloads/ArtificialIntelligencewithPython_ColorImages.pdf.

Errata

Although we have taken every care to ensure the accuracy of our content, mistakes do happen. If you find a mistake in one of our books-maybe a mistake in the text or the code-we would be grateful if you could report this to us. By doing so, you can save other readers from frustration and help us improve subsequent versions of this book. If you find any errata, please report them by visiting http://www.packtpub.com/submit-errata, selecting your book, clicking on the **Errata Submission Form** link, and entering the details of your errata. Once your errata are verified, your submission will be accepted and the errata will be uploaded to our website or added to any list of existing errata under the Errata section of that title.

To view the previously submitted errata, go to https://www.packtpub.com/books/content/support and enter the name of the book in the search field. The required information will appear under the **Errata** section.

Piracy

Piracy of copyrighted material on the Internet is an ongoing problem across all media. At Packt, we take the protection of our copyright and licenses very seriously. If you come across any illegal copies of our works in any form on the Internet, please provide us with the location address or website name immediately so that we can pursue a remedy.

Please contact us at copyright@packtpub.com with a link to the suspected pirated material.

We appreciate your help in protecting our authors and our ability to bring you valuable content.

Questions

If you have a problem with any aspect of this book, you can contact us at questions@packtpub.com, and we will do our best to address the problem.

1
Introduction to Artificial Intelligence

In this chapter, we are going to discuss the concept of **Artificial Intelligence** (**AI**) and how it's applied in the real world. We spend a significant portion of our everyday life interacting with smart systems. It can be in the form of searching for something on the internet, Biometric face recognition, or converting spoken words to text. Artificial Intelligence is at the heart of all this and it's becoming an important part of our modern lifestyle. All these system are complex real-world applications and Artificial Intelligence solves these problems with mathematics and algorithms. During the course of this book, we will learn the fundamental principles that are used to build such applications and then implement them as well. Our overarching goal is to enable you to take up new and challenging Artificial Intelligence problems that you might encounter in your everyday life.

By the end of this chapter, you will know:

- What is AI and why do we need to study it?
- Applications of AI
- Branches of AI
- Turing test
- Rational agents

- General Problem Solvers
- Building an intelligent agent
- Installing Python 3 on various operating systems
- Installing the necessary Python packages

What is Artificial Intelligence?

Artificial Intelligence (**AI**) is a way to make machines think and behave intelligently. These machines are controlled by software inside them, so AI has a lot to do with intelligent software programs that control these machines. It is a science of finding theories and methodologies that can help machines understand the world and accordingly react to situations in the same way that humans do.

If we look closely at how the field of AI has emerged over the last couple of decades, you will see that different researchers tend to focus on different concepts to define AI. In the modern world, AI is used across many verticals in many different forms. We want the machines to sense, reason, think, and act. We want our machines to be rational too.

AI is closely related to the study of human brain. Researchers believe that AI can be accomplished by understanding how the human brain works. By mimicking the way the human brain learns, thinks, and takes action, we can build a machine that can do the same. This can be used as a platform to develop intelligent systems that are capable of learning.

Why do we need to study AI?

AI has the ability to impact every aspect of our lives. The field of AI tries to understand patterns and behaviors of entities. With AI, we want to build smart systems and understand the concept of intelligence as well. The intelligent systems that we construct are very useful in understanding how an intelligent system like our brain goes about constructing another intelligent system.

Let's take a look at how our brain processes information:

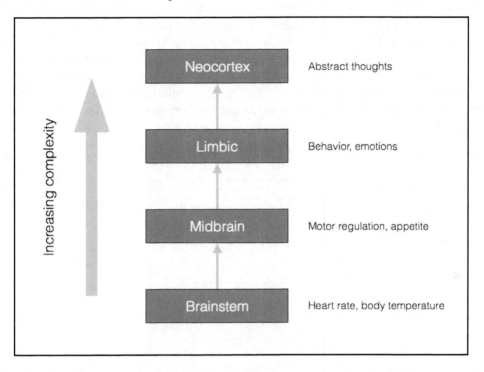

Compared to some other fields such as Mathematics or Physics that have been around for centuries, AI is relatively in its infancy. Over the last couple of decades, AI has produced some spectacular products such as self-driving cars and intelligent robots that can walk. Based on the direction in which we are heading, it's pretty obvious that achieving intelligence will have a great impact on our lives in the coming years.

Introduction to Artificial Intelligence

We can't help but wonder how the human brain manages to do so much with such effortless ease. We can recognize objects, understand languages, learn new things, and perform many more sophisticated tasks with our brain. How does the human brain do this? When you try to do this with a machine, you will see that it falls way behind! For example, when we try to look for things such as extraterrestrial life or time travel, we don't know if those things exist. The good thing about the holy grail of AI is that we know it exists. Our brain is the holy grail! It is a spectacular example of an intelligent system. All we have to do is to mimic its functionality to create an intelligent system that can do something similar, possibly even more.

Let's see how raw data gets converted to wisdom through various levels of processing:

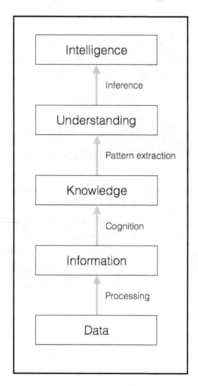

One of the main reasons we want to study AI is to automate many things. We live in a world where:

- We deal with huge and insurmountable amounts of data. The human brain can't keep track of so much data.
- Data originates from multiple sources simultaneously.
- The data is unorganized and chaotic.
- Knowledge derived from this data has to be updated constantly because the data itself keeps changing.
- The sensing and actuation has to happen in real time with high precision.

Even though the human brain is great at analyzing things around us, it cannot keep up with the preceding conditions. Hence, we need to design and develop intelligent machines that can do this. We need AI systems that can:

- Handle large amounts of data in an efficient way. With the advent of Cloud Computing, we are now able to store huge amounts of data.
- Ingest data simultaneously from multiple sources without any lag.
- Index and organize data in a way that allows us to derive insights.
- Learn from new data and update constantly using the right learning algorithms.
- Think and respond to situations based on the conditions in real time.

AI techniques are actively being used to make existing machines smarter, so that they can execute faster and more efficiently.

Applications of AI

Now that we know how information gets processed, let's see where AI appears in the real world. AI manifests itself in various different forms across multiple fields, so it's important to understand how it's useful in various domains. AI has been used across many industries and it continues to expand rapidly. Some of the most popular areas include:

- **Computer Vision**: These are the systems that deal with visual data such as images and videos. These systems understand the content and extract insights based on the use case. For example, Google uses reverse image search to search for visually similar images across the Web.

- **Natural Language Processing**: This field deals with understanding text. We can interact with a machine by typing natural language sentences. Search engines use this extensively to deliver the right search results.
- **Speech Recognition**: These systems are capable of hearing and understanding spoken words. For example, there are intelligent personal assistants on our smartphones that can understand what we are saying and give relevant information or perform an action based on that.
- **Expert Systems**: These systems use AI techniques to provide advice or make decisions. They usually use databases of expert knowledge areas such as finance, medicine, marketing, and so on to give advice about what to do next. Let's see what an expert system looks like and how it interacts with the user:

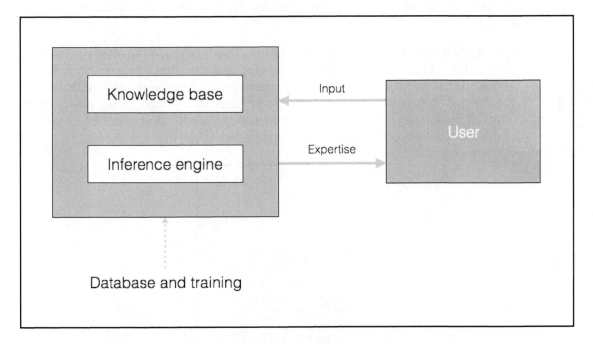

- **Games**: AI is used extensively in the gaming industry. It is used to design intelligent agents that can compete with humans. For example, **AlphaGo** is a computer program that can play the strategy game Go. It is also used in designing many other types of games where we expect the computer to behave intelligently.

- **Robotics**: Robotic systems actually combine many concepts in AI. These systems are able to perform many different tasks. Depending on the situation, robots have sensors and actuators that can do different things. These sensors can see things in front of them and measure the temperature, heat, movements, and so on. They have processors on board that compute various things in real time. They are also capable of adapting to the new environments.

Branches of AI

It is important to understand the various fields of study within AI so that we can choose the right framework to solve a given real-world problem. Here's a list of topics that are dominant:

- **Machine learning and pattern recognition**: This is perhaps the most popular form of AI out there. We design and develop software that can learn from data. Based on these learning models, we perform predictions on unknown data. One of the main constraints here is that these programs are limited to the power of the data. If the dataset is small, then the learning models would be limited as well. Let's see what a typical machine learning system looks like:

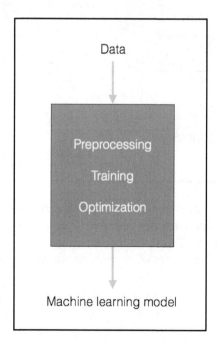

Chapter 1

When a system makes an observation, it is trained to compare it with what it has already seen in the form of a pattern. For example, in a face recognition system, the software will try to match the pattern of eyes, nose, lips, eyebrows, and so on in order to find a face in the existing database of users.

- **Logic-based AI**: Mathematical logic is used to execute computer programs in logic-based AI. A program written in logic-based AI is basically a set of statements in logical form that express facts and rules about a particular problem domain. This is used extensively in pattern matching, language parsing, semantic analysis, and so on.
- **Search**: The Search techniques are used extensively in AI programs. These programs examine a large number of possibilities and then pick the most optimal path. For example, this is used a lot in strategy games such as Chess, networking, resource allocation, scheduling, and so on.
- **Knowledge representation**: The facts about the world around us need to be represented in some way for a system to make sense of them. The languages of mathematical logic are frequently used here. If knowledge is represented efficiently, systems can be smarter and more intelligent. Ontology is a closely related field of study that deals with the kinds of objects that exist. It is a formal definition of the properties and relationships of the entities that exist in a particular domain. This is usually done with a particular taxonomy or a hierarchical structure of some kind. The following diagram shows the difference between information and knowledge:

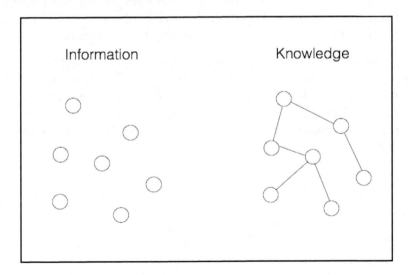

- **Planning**: This field deals with optimal planning that gives us maximum returns with minimal costs. These software programs start with facts about the particular situation and a statement of a goal. These programs are also aware of the facts of the world, so that they know what the rules are. From this information, they generate the most optimal plan to achieve the goal.
- **Heuristics**: A heuristic is a technique used to solve a given problem that's practical and useful in solving the problem in the short term, but not guaranteed to be optimal. This is more like an educated guess on what approach we should take to solve a problem. In AI, we frequently encounter situations where we cannot check every single possibility to pick the best option. So we need to use heuristics to achieve the goal. They are used extensively in AI in fields such as robotics, search engines, and so on.
- **Genetic programming**: Genetic programming is a way to get programs to solve a task, by mating programs and selecting the fittest. The programs are encoded as a set of genes, using an algorithm to get a program that is able to perform the given task really well.

Defining intelligence using Turing Test

The legendary computer scientist and mathematician, *Alan Turing*, proposed the Turing Test to provide a definition of intelligence. It is a test to see if a computer can learn to mimic human behavior. He defined intelligent behavior as the ability to achieve human-level intelligence during a conversation. This performance should be sufficient to trick an interrogator into thinking that the answers are coming from a human.

To see if a machine can do this, he proposed a test setup: he proposed that a human should interrogate the machine through a text interface. Another constraint is that the human cannot know who's on the other side of the interrogation, which means it can either be a machine or a human. To enable this setup, a human will be interacting with two entities through a text interface. These two entities are called respondents. One of them will be a human and the other one will be the machine.

The respondent machine passes the test if the interrogator is unable to tell whether the answers are coming from a machine or a human. The following diagram shows the setup of a Turing Test:

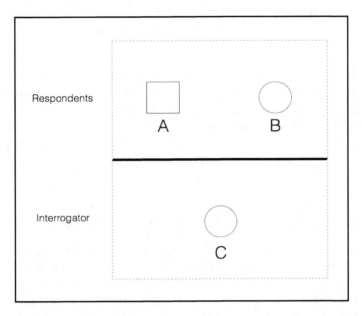

As you can imagine, this is quite a difficult task for the respondent machine. There are a lot of things going on during a conversation. At the very minimum, the machine needs to be well versed with the following things:

- **Natural Language Processing**: The machine needs this to communicate with the interrogator. The machine needs to parse the sentence, extract the context, and give an appropriate answer.
- **Knowledge Representation**: The machine needs to store the information provided before the interrogation. It also needs to keep track of the information being provided during the conversation so that it can respond appropriately if it comes up again.

- **Reasoning**: It's important for the machine to understand how to interpret the information that gets stored. Humans tend to do this automatically to draw conclusions in real time.
- **Machine Learning**: This is needed so that the machine can adapt to new conditions in real time. The machine needs to analyze and detect patterns so that it can draw inferences.

You must be wondering why the human is communicating with a text interface. According to Turing, physical simulation of a person is unnecessary for intelligence. That's the reason the Turing Test avoids direct physical interaction between the human and the machine. There is another thing called the Total Turing Test that deals with vision and movement. To pass this test, the machine needs to see objects using computer vision and move around using Robotics.

Making machines think like humans

For decades, we have been trying to get the machine to think like a human. In order to make this happen, we need to understand how humans think in the first place. How do we understand the nature of human thinking? One way to do this would be to note down how we respond to things. But this quickly becomes intractable, because there are too many things to note down. Another way to do this is to conduct an experiment based on a predefined format. We develop a certain number of questions to encompass a wide variety of human topics, and then see how people respond to it.

Once we gather enough data, we can create a model to simulate the human process. This model can be used to create software that can think like humans. Of course this is easier said than done! All we care about is the output of the program given a particular input. If the program behaves in a way that matches human behavior, then we can say that humans have a similar thinking mechanism.

The following diagram shows different levels of thinking and how our brain prioritizes things:

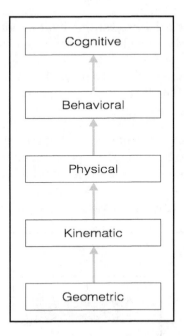

Within computer science, there is a field of study called **Cognitive Modeling** that deals with simulating the human thinking process. It tries to understand how humans solve problems. It takes the mental processes that go into this problem solving process and turns it into a software model. This model can then be used to simulate human behavior. Cognitive modeling is used in a variety of AI applications such as deep learning, expert systems, Natural Language Processing, robotics, and so on.

Building rational agents

A lot of research in AI is focused on building rational agents. What exactly is a rational agent? Before that, let us define the word rationality. Rationality refers to doing the right thing in a given circumstance. This needs to be performed in such a way that there is maximum benefit to the entity performing the action. An agent is said to act rationally if, given a set of rules, it takes actions to achieve its goals. It just perceives and acts according to the information that's available. This system is used a lot in AI to design robots when they are sent to navigate unknown terrains.

How do we define the *right* thing? The answer is that it depends on the objectives of the agent. The agent is supposed to be intelligent and independent. We want to impart the ability to adapt to new situations. It should understand its environment and then act accordingly to achieve an outcome that is in its best interests. The best interests are dictated by the overall goal it wants to achieve. Let's see how an input gets converted to action:

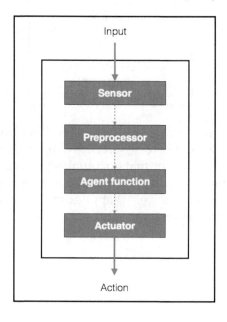

How do we define the performance measure for a rational agent? One might say that it is directly proportional to the degree of success. The agent is set up to achieve a particular task, so the performance measure depends on what percentage of that task is complete. But we must think as to what constitutes rationality in its entirety. If it's just about results, can the agent take any action to get there?

Making the right inferences is definitely a part of being rational, because the agent has to act rationally to achieve its goals. This will help it draw conclusions that can be used successively. What about situations where there are no provably right things to do? There are situations where the agent doesn't know what to do, but it still has to do something. In this situation, we cannot include the concept of inference to define rational behavior.

General Problem Solver

The **General Problem Solver** (**GPS**) was an AI program proposed by *HerbertSimon, J.C. Shaw*, and *Allen Newell*. It was the first useful computer program that came into existence in the AI world. The goal was to make it work as a universal problem-solving machine. Of course there were many software programs that existed before, but these programs performed specific tasks. GPS was the first program that was intended to solve any general problem. GPS was supposed to solve all the problems using the same base algorithm for every problem.

As you must have realized, this is quite an uphill battle! To program the GPS, the authors created a new language called **Information Processing Language** (**IPL**). The basic premise is to express any problem with a set of well-formed formulas. These formulas would be a part of a directed graph with multiple sources and sinks. In a graph, the source refers to the starting node and the sink refers to the ending node. In the case of GPS, the source refers to axioms and the sink refers to the conclusions.

Even though GPS was intended to be a general purpose, it could only solve well-defined problems, such as proving mathematical theorems in geometry and logic. It could also solve word puzzles and play chess. The reason was that these problems could be formalized to a reasonable extent. But in the real world, this quickly becomes intractable because of the number of possible paths you can take. If it tries to brute force a problem by counting the number of walks in a graph, it becomes computationally infeasible.

Solving a problem with GPS

Let's see how to structure a given problem to solve it using GPS:

1. The first step is to define the goals. Let's say our goal is to get some milk from the grocery store.
2. The next step is to define the preconditions. These preconditions are in reference to the goals. To get milk from the grocery store, we need to have a mode of transportation and the grocery store should have milk available.
3. After this, we need to define the operators. If my mode of transportation is a car and if the car is low on fuel, then we need to ensure that we can pay the fueling station. We need to ensure that you can pay for the milk at the store.

An operator takes care of the conditions and everything that affects them. It consists of actions, preconditions, and the changes resulting from taking actions. In this case, the action is giving money to the grocery store. Of course, this is contingent upon you having the money in the first place, which is the precondition. By giving them the money, you are changing your money condition, which will result in you getting the milk.

GPS will work as long as you can frame the problem like we did just now. The constraint is that it uses the search process to perform its job, which is way too computationally complex and time consuming for any meaningful real-world application.

Building an intelligent agent

There are many ways to impart intelligence to an agent. The most commonly used techniques include machine learning, stored knowledge, rules, and so on. In this section, we will focus on machine learning. In this method, the way we impart intelligence to an agent is through data and training.

Let's see how an intelligent agent interacts with the environment:

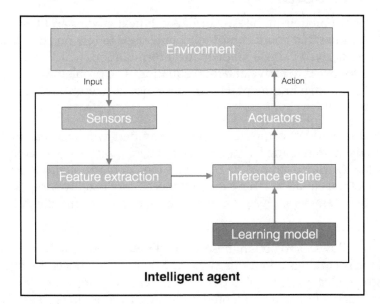

With machine learning, we want to program our machines to use labeled data to solve a given problem. By going through the data and the associated labels, the machine learns how to extract patterns and relationships.

In the preceding example, the intelligent agent depends on the learning model to run the inference engine. Once the sensor perceives the input, it sends it to the feature extraction block. Once the relevant features are extracted, the trained inference engine performs a prediction based on the learning model. This learning model is built using machine learning. The inference engine then takes a decision and sends it to the actuator, which then takes the required action in the real world.

There are many applications of machine learning that exist today. It is used in image recognition, robotics, speech recognition, predicting stock market behavior, and so on. In order to understand machine learning and build a complete solution, you will have to be familiar with many techniques from different fields such as pattern recognition, artificial neural networks, data mining, statistics, and so on.

Types of models

There are two types of models in the AI world: Analytical models and Learned models. Before we had machines that could compute, people used to rely on analytical models. These models were derived using a mathematical formulation, which is basically a sequence of steps followed to arrive at a final equation. The problem with this approach is that it was based on human judgment. Hence these models were simplistic and inaccurate with just a few parameters.

We then entered the world of computers. These computers were good at analyzing data. So, people increasingly started using learned models. These models are obtained through the process of training. During training, the machines look at many examples of inputs and outputs to arrive at the equation. These learned models are usually complex and accurate, with thousands of parameters. This gives rise to a very complex mathematical equation that governs the data.

Machine Learning allows us to obtain these learned models that can be used in an inference engine. One of the best things about this is the fact that we don't need to derive the underlying mathematical formula. You don't need to know complex mathematics, because the machine derives the formula based on data. All we need to do is create the list of inputs and the corresponding outputs. The learned model that we get is just the relationship between labeled inputs and the desired outputs.

Installing Python 3

We will be using Python 3 throughout this book. Make sure you have installed the latest version of Python 3 on your machine. Type the following command on your Terminal to check:

```
$ python3 --version
```

If you see something like Python 3.x.x (where x.x are version numbers) printed on your terminal, you are good to go. If not, installing it is pretty straightforward.

Installing on Ubuntu

Python 3 is already installed by default on Ubuntu 14.xx and above. If not, you can install it using the following command:

```
$ sudo apt-get install python3
```

Run the check command like we did earlier:

```
$ python3 --version
```

You should see the version number printed on your Terminal.

Installing on Mac OS X

If you are on Mac OS X, it is recommended that you use Homebrew to install Python 3. It is a great package installer for Mac OS X and it is really easy to use. If you don't have Homebrew, you can install it using the following command:

```
$ ruby -e "$(curl -fsSL https://raw.githubusercontent.com/Homebrew/install/master/install)"
```

Let's update the package manager:

```
$ brew update
```

Let's install Python 3:

```
$ brew install python3
```

Run the check command like we did earlier:

```
$ python3 --version
```

You should see the version number printed on your Terminal.

Installing on Windows

If you use Windows, it is recommended that you use a `SciPy-stack` compatible distribution of Python 3. Anaconda is pretty popular and easy to use. You can find the installation instructions at: `https://www.continuum.io/downloads`.

If you want to check out other SciPy-stack compatible distributions of Python 3, you can find them at `http://www.scipy.org/install.html`. The good part about these distributions is that they come with all the necessary packages preinstalled. If you use one of these versions, you don't need to install the packages separately.

Once you install it, run the check command like we did earlier:

```
$ python3 --version
```

You should see the version number printed on your Terminal.

Installing packages

During the course of this book, we will use various packages such as NumPy, SciPy, scikit-learn, and matplotlib. Make sure you install these packages before you proceed.

If you use Ubuntu or Mac OS X, installing these packages is pretty straightforward. All these packages can be installed using a one-line command on the terminal. Here are the relevant links for installation:

- **NumPy**: `http://docs.scipy.org/doc/numpy-1.10.1/user/install.html`
- **SciPy**: `http://www.scipy.org/install.html`
- **scikit-learn**: `http://scikit-learn.org/stable/install.html`
- **matplotlib**: `http://matplotlib.org/1.4.2/users/installing.html`

If you are on Windows, you should have installed a `SciPy-stack` compatible version of Python 3.

Loading data

In order to build a learning model, we need data that's representative of the world. Now that we have installed the necessary Python packages, let's see how to use the packages to interact with data. Go into the Python terminal by typing the following command:

```
$ python3
```

Let's import the package containing all the datasets:

```
>>> from sklearn import datasets
```

Let's load the house prices dataset:

```
>>> house_prices = datasets.load_boston()
```

Print the data:

```
>>> print(house_prices.data)
```

You will see an output like this printed on your Terminal:

```
>>> print(house_prices.data)
[[  6.32000000e-03   1.80000000e+01   2.31000000e+00 ...,   1.53000000e+01
    3.96900000e+02   4.98000000e+00]
 [  2.73100000e-02   0.00000000e+00   7.07000000e+00 ...,   1.78000000e+01
    3.96900000e+02   9.14000000e+00]
 [  2.72900000e-02   0.00000000e+00   7.07000000e+00 ...,   1.78000000e+01
    3.92830000e+02   4.03000000e+00]
 ...,
 [  6.07600000e-02   0.00000000e+00   1.19300000e+01 ...,   2.10000000e+01
    3.96900000e+02   5.64000000e+00]
 [  1.09590000e-01   0.00000000e+00   1.19300000e+01 ...,   2.10000000e+01
    3.93450000e+02   6.48000000e+00]
 [  4.74100000e-02   0.00000000e+00   1.19300000e+01 ...,   2.10000000e+01
    3.96900000e+02   7.88000000e+00]]
```

Introduction to Artificial Intelligence

Let's check out the labels:

You will see the following printed on your Terminal:

```
>>> print(house_prices.target)
[ 24.   21.6  34.7  33.4  36.2  28.7  22.9  27.1  16.5  18.9  15.   18.9
  21.7  20.4  18.2  19.9  23.1  17.5  20.2  18.2  13.6  19.6  15.2  14.5
  15.6  13.9  16.6  14.8  18.4  21.   12.7  14.5  13.2  13.1  13.5  18.9
  20.   21.   24.7  30.8  34.9  26.6  25.3  24.7  21.2  19.3  20.   16.6
  14.4  19.4  19.7  20.5  25.   23.4  18.9  35.4  24.7  31.6  23.3  19.6
  18.7  16.   22.2  25.   33.   23.5  19.4  22.   17.4  20.9  24.2  21.7
  22.8  23.4  24.1  21.4  20.   20.8  21.2  20.3  28.   23.9  24.8  22.9
  23.9  26.6  22.5  22.2  23.6  28.7  22.6  22.   22.9  25.   20.6  28.4
  21.4  38.7  43.8  33.2  27.5  26.5  18.6  19.3  20.1  19.5  19.5  20.4
  19.8  19.4  21.7  22.8  18.8  18.7  18.5  18.3  21.2  19.2  20.4  19.3
  22.   20.3  20.5  17.3  18.8  21.4  15.7  16.2  18.   14.3  19.2  19.6
  23.   18.4  15.6  18.1  17.4  17.1  13.3  17.8  14.   14.4  13.4  15.6
  11.8  13.8  15.6  14.6  17.8  15.4  21.5  19.6  15.3  19.4  17.   15.6
  13.1  41.3  24.3  23.3  27.   50.   50.   50.   22.7  25.   50.   23.8
  23.8  22.3  17.4  19.1  23.1  23.6  22.6  29.4  23.2  24.6  29.9  37.2
  39.8  36.2  37.9  32.5  26.4  29.6  50.   32.   29.8  34.9  37.   30.5
  36.4  31.1  29.1  50.   33.3  30.3  34.6  34.9  32.9  24.1  42.3  48.5
  50.   22.6  24.4  22.5  24.4  20.   21.7  19.3  22.4  28.1  23.7  25.
  23.3  28.7  21.5  23.   26.7  21.7  27.5  30.1  44.8  50.   37.6  31.6
  46.7  31.5  24.3  31.7  41.7  48.3  29.   24.   25.1  31.5  23.7  23.3
```

The actual array is larger, so the image represents the first few values in that array.

There are also image datasets available in the scikit-learn package. Each image is of shape 8×8. Let's load it:

```
>>> digits = datasets.load_digits()
```

Print the fifth image:

```
>>> print(digits.images[4])
```

[28]

You will see the following on your Terminal:

```
>>> print(digits.images[4])
[[  0.   0.   0.   1.  11.   0.   0.   0.]
 [  0.   0.   0.   7.   8.   0.   0.   0.]
 [  0.   0.   1.  13.   6.   2.   2.   0.]
 [  0.   0.   7.  15.   0.   9.   8.   0.]
 [  0.   5.  16.  10.   0.  16.   6.   0.]
 [  0.   4.  15.  16.  13.  16.   1.   0.]
 [  0.   0.   0.   3.  15.  10.   0.   0.]
 [  0.   0.   0.   2.  16.   4.   0.   0.]]
```

As you can see, it has eight rows and eight columns.

Summary

In this chapter, we learned what AI is all about and why we need to study it. We discussed various applications and branches of AI. We understood what the Turing test is and how it's conducted. We learned how to make machines think like humans. We discussed the concept of rational agents and how they should be designed. We learned about General Problem Solver (GPS) and how to solve a problem using GPS. We discussed how to develop an intelligent agent using machine learning. We covered different types of models as well.

We discussed how to install Python 3 on various operating systems. We learned how to install the necessary packages required to build AI applications. We discussed how to use the packages to load data that's available in scikit-learn. In the next chapter, we will learn about supervised learning and how to build models for classification and regression.

2
Classification and Regression Using Supervised Learning

In this chapter, we are going to learn about classification and regression of data using supervised learning techniques. By the end of this chapter, you will know about these topics:

- What is the difference between supervised and unsupervised learning?
- What is classification?
- How to preprocess data using various methods
- What is label encoding?
- How to build a logistic regression classifier
- What is Naïve Bayes classifier?
- What is a confusion matrix?
- What are Support Vector Machines and how to build a classifier based on that?
- What is linear and polynomial regression?
- How to build a linear regressor for single variable and multivariable data
- How to estimate housing prices using Support Vector Regressor

Supervised versus unsupervised learning

One of the most common ways to impart artificial intelligence into a machine is through machine learning. The world of machine learning is broadly divided into supervised and unsupervised learning. There are other divisions too, but we'll discuss those later.

Supervised learning refers to the process of building a machine learning model that is based on labeled training data. For example, let's say that we want to build a system to automatically predict the income of a person, based on various parameters such as age, education, location, and so on. To do this, we need to create a database of people with all the necessary details and label it. By doing this, we are telling our algorithm what parameters correspond to what income. Based on this mapping, the algorithm will learn how to calculate the income of a person using the parameters provided to it.

Unsupervised learning refers to the process of building a machine learning model without relying on labeled training data. In some sense, it is the opposite of what we just discussed in the previous paragraph. Since there are no labels available, you need to extract insights based on just the data given to you. For example, let's say that we want to build a system where we have to separate a set of data points into multiple groups. The tricky thing here is that we don't know exactly what the criteria of separation should be. Hence, an unsupervised learning algorithm needs to separate the given dataset into a number of groups in the best way possible.

What is classification?

In this chapter, we will discuss supervised classification techniques. The process of classification is one such technique where we classify data into a given number of classes. During classification, we arrange data into a fixed number of categories so that it can be used most effectively and efficiently.

In machine learning, classification solves the problem of identifying the category to which a new data point belongs. We build the classification model based on the training dataset containing data points and the corresponding labels. For example, let's say that we want to check whether the given image contains a person's face or not. We would build a training dataset containing classes corresponding to these two classes: `face` and `no-face`. We then train the model based on the training samples we have. This trained model is then used for inference.

A good classification system makes it easy to find and retrieve data. This is used extensively in face recognition, spam identification, recommendation engines, and so on. The algorithms for data classification will come up with the right criteria to separate the given data into the given number of classes.

We need to provide a sufficiently large number of samples so that it can generalize those criteria. If there is an insufficient number of samples, then the algorithm will overfit to the training data. This means that it won't perform well on unknown data because it fine-tuned the model too much to fit into the patterns observed in training data. This is actually a very common problem that occurs in the world of machine learning. It's good to consider this factor when you build various machine learning models.

Preprocessing data

We deal with a lot of raw data in the real world. Machine learning algorithms expect data to be formatted in a certain way before they start the training process. In order to prepare the data for ingestion by machine learning algorithms, we have to preprocess it and convert it into the right format. Let's see how to do it.

Create a new Python file and import the following packages:

```
import numpy as np
from sklearn import preprocessing
```

Let's define some sample data:

```
input_data = np.array([[5.1, -2.9, 3.3],
                       [-1.2, 7.8, -6.1],
                       [3.9, 0.4, 2.1],
                       [7.3, -9.9, -4.5]])
```

We will be talking about several different preprocessing techniques. Let's start with binarization:

- Binarization
- Mean removal
- Scaling
- Normalization

Let's take a look at each technique, starting with the first.

Binarization

This process is used when we want to convert our numerical values into boolean values. Let's use an inbuilt method to binarize input data using 2.1 as the threshold value.

Add the following lines to the same Python file:

```
# Binarize data
data_binarized =
preprocessing.Binarizer(threshold=2.1).transform(input_data)
print("\nBinarized data:\n", data_binarized)
```

If you run the code, you will see the following output:

```
Binarized data:
 [[ 1.  0.  1.]
  [ 0.  1.  0.]
  [ 1.  0.  0.]
  [ 1.  0.  0.]]
```

As we can see here, all the values above 2.1 become 1. The remaining values become 0.

Mean removal

Removing the mean is a common preprocessing technique used in machine learning. It's usually useful to remove the mean from our feature vector, so that each feature is centered on zero. We do this in order to remove bias from the features in our feature vector.

Add the following lines to the same Python file as in the previous section:

```
# Print mean and standard deviation
print("\nBEFORE:")
print("Mean =", input_data.mean(axis=0))
print("Std deviation =", input_data.std(axis=0))
```

The preceding line displays the mean and standard deviation of the input data. Let's remove the mean:

```
# Remove mean
data_scaled = preprocessing.scale(input_data)
print("\nAFTER:")
print("Mean =", data_scaled.mean(axis=0))
print("Std deviation =", data_scaled.std(axis=0))
```

If you run the code, you will see the following printed on your Terminal:

```
BEFORE:
Mean = [ 3.775 -1.15  -1.3  ]
Std deviation = [ 3.12039661  6.36651396  4.0620192 ]
AFTER:
Mean = [   1.11022302e-16   0.00000000e+00   2.77555756e-17]
Std deviation = [ 1.  1.  1.]
```

As seen from the values obtained, the mean value is very close to 0 and standard deviation is 1.

Scaling

In our feature vector, the value of each feature can vary between many random values. So it becomes important to scale those features so that it is a level playing field for the machine learning algorithm to train on. We don't want any feature to be artificially large or small just because of the nature of the measurements.

Add the following line to the same Python file:

```
# Min max scaling
data_scaler_minmax = preprocessing.MinMaxScaler(feature_range=(0, 1))
data_scaled_minmax = data_scaler_minmax.fit_transform(input_data)
print("\nMin max scaled data:\n", data_scaled_minmax)
```

If you run the code, you will see the following printed on your Terminal:

```
Min max scaled data:
 [[ 0.74117647  0.39548023  1.        ]
  [ 0.          1.          0.        ]
  [ 0.6         0.5819209   0.87234043]
  [ 1.          0.          0.17021277]]
```

Each row is scaled so that the maximum value is 1 and all the other values are relative to this value.

Normalization

We use the process of normalization to modify the values in the feature vector so that we can measure them on a common scale. In machine learning, we use many different forms of normalization. Some of the most common forms of normalization aim to modify the values so that they sum up to 1. **L1 normalization**, which refers to **Least Absolute Deviations**, works by making sure that the sum of absolute values is *1* in each row. **L2 normalization**, which refers to least squares, works by making sure that the sum of squares is *1*.

In general, L1 normalization technique is considered more robust than L2 normalization technique. L1 normalization technique is robust because it is resistant to outliers in the data. A lot of times, data tends to contain outliers and we cannot do anything about it. We want to use techniques that can safely and effectively ignore them during the calculations. If we are solving a problem where outliers are important, then maybe L2 normalization becomes a better choice.

Add the following lines to the same Python file:

```python
# Normalize data
data_normalized_l1 = preprocessing.normalize(input_data, norm='l1')
data_normalized_l2 = preprocessing.normalize(input_data, norm='l2')
print("\nL1 normalized data:\n", data_normalized_l1)
print("\nL2 normalized data:\n", data_normalized_l2)
```

If you run the code, you will see the following printed on your Terminal:

```
L1 normalized data:
[[ 0.45132743 -0.25663717  0.2920354 ]
 [-0.0794702   0.51655629 -0.40397351]
 [ 0.609375    0.0625      0.328125  ]
 [ 0.33640553 -0.4562212  -0.20737327]]
L2 normalized data:
[[ 0.75765788 -0.43082507  0.49024922]
 [-0.12030718  0.78199664 -0.61156148]
 [ 0.87690281  0.08993875  0.47217844]
 [ 0.55734935 -0.75585734 -0.34357152]]
```

The code for this entire section is given in the `preprocessing.py` file.

Label encoding

When we perform classification, we usually deal with a lot of labels. These labels can be in the form of words, numbers, or something else. The machine learning functions in **sklearn** expect them to be numbers. So if they are already numbers, then we can use them directly to start training. But this is not usually the case.

In the real world, labels are in the form of words, because words are human readable. We label our training data with words so that the mapping can be tracked. To convert word labels into numbers, we need to use a label encoder. Label encoding refers to the process of transforming the word labels into numerical form. This enables the algorithms to operate on our data.

Create a new Python file and import the following packages:

```
import numpy as np
from sklearn import preprocessing
```

Define some sample labels:

```
# Sample input labels
input_labels = ['red', 'black', 'red', 'green', 'black', 'yellow', 'white']
```

Create the label encoder object and train it:

```
# Create label encoder and fit the labels
encoder = preprocessing.LabelEncoder()
encoder.fit(input_labels)
```

Print the mapping between words and numbers:

```
# Print the mapping
print("\nLabel mapping:")
for i, item in enumerate(encoder.classes_):
    print(item, '-->', i)
```

Let's encode a set of randomly ordered labels to see how it performs:

```
# Encode a set of labels using the encoder
test_labels = ['green', 'red', 'black']
encoded_values = encoder.transform(test_labels)
print("\nLabels =", test_labels)
print("Encoded values =", list(encoded_values))
```

Let's decode a random set of numbers:

```
# Decode a set of values using the encoder
encoded_values = [3, 0, 4, 1]
decoded_list = encoder.inverse_transform(encoded_values)
print("\nEncoded values =", encoded_values)
print("Decoded labels =", list(decoded_list))
```

If you run the code, you will see the following output:

```
Label mapping:
black --> 0
green --> 1
red --> 2
white --> 3
yellow --> 4

Labels = ['green', 'red', 'black']
Encoded values = [1, 2, 0]

Encoded values = [3, 0, 4, 1]
Decoded labels = ['white', 'black', 'yellow', 'green']
```

You can check the mapping to see that the encoding and decoding steps are correct. The code for this section is given in the `label_encoder.py` file.

Logistic Regression classifier

Logistic regression is a technique that is used to explain the relationship between input variables and output variables. The input variables are assumed to be independent and the output variable is referred to as the dependent variable. The dependent variable can take only a fixed set of values. These values correspond to the classes of the classification problem.

Our goal is to identify the relationship between the independent variables and the dependent variables by estimating the probabilities using a logistic function. This logistic function is a **sigmoid curve** that's used to build the function with various parameters. It is very closely related to generalized linear model analysis, where we try to fit a line to a bunch of points to minimize the error. Instead of using linear regression, we use logistic regression. Logistic regression by itself is actually not a classification technique, but we use it in this way so as to facilitate classification. It is used very commonly in machine learning because of its simplicity. Let's see how to build a classifier using logistic regression. Make sure you have `Tkinter` package installed on your system before you proceed. If you don't, you can find it at: https://docs.python.org/2/library/tkinter.html.

Create a new Python file and import the following packages. We will be importing a function from the file `utilities.py`. We will be looking into that function very soon. But for now, let's import it:

```
import numpy as np
from sklearn import linear_model
import matplotlib.pyplot as plt

from utilities import visualize_classifier
```

Define sample input data with two-dimensional vectors and corresponding labels:

```
# Define sample input data
X = np.array([[3.1, 7.2], [4, 6.7], [2.9, 8], [5.1, 4.5], [6, 5], [5.6, 5],
    [3.3, 0.4], [3.9, 0.9], [2.8, 1], [0.5, 3.4], [1, 4], [0.6, 4.9]])
y = np.array([0, 0, 0, 1, 1, 1, 2, 2, 2, 3, 3, 3])
```

We will train the classifier using this labeled data. Now create the logistic regression classifier object:

```
# Create the logistic regression classifier
classifier = linear_model.LogisticRegression(solver='liblinear', C=1)
```

Train the classifier using the data that we defined earlier:

```
# Train the classifier
classifier.fit(X, y)
```

Visualize the performance of the classifier by looking at the boundaries of the classes:

```
# Visualize the performance of the classifier
visualize_classifier(classifier, X, y)
```

We need to define this function before we can use it. We will be using this multiple times in this chapter, so it's better to define it in a separate file and import the function. This function is given in the `utilities.py` file provided to you.

Create a new Python file and import the following packages:

```
import numpy as np
import matplotlib.pyplot as plt
```

Create the function definition by taking the classifier object, input data, and labels as input parameters:

```
def visualize_classifier(classifier, X, y):
    # Define the minimum and maximum values for X and Y
    # that will be used in the mesh grid
    min_x, max_x = X[:, 0].min() - 1.0, X[:, 0].max() + 1.0
    min_y, max_y = X[:, 1].min() - 1.0, X[:, 1].max() + 1.0
```

We also defined the minimum and maximum values of X and Y directions that will be used in our mesh grid. This grid is basically a set of values that is used to evaluate the function, so that we can visualize the boundaries of the classes. Define the step size for the grid and create it using the minimum and maximum values:

```
    # Define the step size to use in plotting the mesh grid
    mesh_step_size = 0.01

    # Define the mesh grid of X and Y values
    x_vals, y_vals = np.meshgrid(np.arange(min_x, max_x, mesh_step_size), np.arange(min_y, max_y, mesh_step_size))
```

Run the classifier on all the points on the grid:

```
    # Run the classifier on the mesh grid
    output = classifier.predict(np.c_[x_vals.ravel(), y_vals.ravel()])

    # Reshape the output array
    output = output.reshape(x_vals.shape)
```

Create the figure, pick a color scheme, and overlay all the points:

```
    # Create a plot
    plt.figure()

    # Choose a color scheme for the plot
    plt.pcolormesh(x_vals, y_vals, output, cmap=plt.cm.gray)

    # Overlay the training points on the plot
    plt.scatter(X[:, 0], X[:, 1], c=y, s=75, edgecolors='black', linewidth=1, cmap=plt.cm.Paired)
```

Specify the boundaries of the plots using the minimum and maximum values, add the tick marks, and display the figure:

```
# Specify the boundaries of the plot
plt.xlim(x_vals.min(), x_vals.max())
plt.ylim(y_vals.min(), y_vals.max())

# Specify the ticks on the X and Y axes
plt.xticks((np.arange(int(X[:, 0].min() - 1), int(X[:, 0].max() + 1), 1.0)))
plt.yticks((np.arange(int(X[:, 1].min() - 1), int(X[:, 1].max() + 1), 1.0)))

plt.show()
```

If you run the code, you will see the following screenshot:

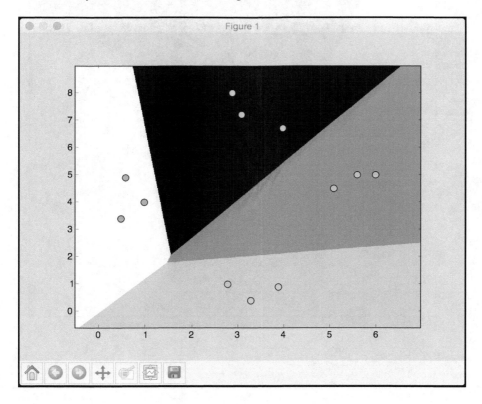

If you change the value of C to *100* in the following line, you will see that the boundaries become more accurate:

```
classifier = linear_model.LogisticRegression(solver='liblinear', C=100)
```

The reason is that C imposes a certain penalty on misclassification, so the algorithm customizes more to the training data. You should be careful with this parameter, because if you increase it by a lot, it will overfit to the training data and it won't generalize well.

If you run the code with C set to 100, you will see the following screenshot:

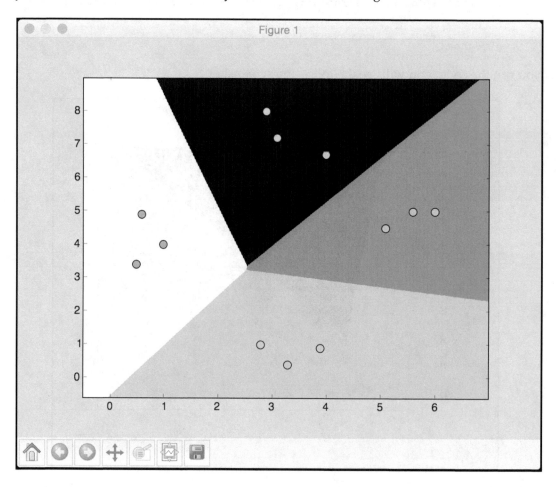

If you compare with the earlier figure, you will see that the boundaries are now better. The code for this section is given in the `logistic_regression.py` file.

Naïve Bayes classifier

Naïve Bayes is a technique used to build classifiers using Bayes theorem. Bayes theorem describes the probability of an event occurring based on different conditions that are related to this event. We build a Naïve Bayes classifier by assigning class labels to problem instances. These problem instances are represented as vectors of feature values. The assumption here is that the value of any given feature is independent of the value of any other feature. This is called the independence assumption, which is the *naïve* part of a Naïve Bayes classifier.

Given the class variable, we can just see how a given feature affects, it regardless of its affect on other features. For example, an animal may be considered a cheetah if it is spotted, has four legs, has a tail, and runs at about *70 MPH*. A Naïve Bayes classifier considers that each of these features contributes independently to the outcome. The outcome refers to the probability that this animal is a cheetah. We don't concern ourselves with the correlations that may exist between skin patterns, number of legs, presence of a tail, and movement speed. Let's see how to build a Naïve Bayes classifier.

Create a new Python file and import the following packages:

```
import numpy as np
import matplotlib.pyplot as plt
from sklearn.Naïve_bayes import GaussianNB
from sklearn import cross_validation

from utilities import visualize_classifier
```

We will be using the file `data_multivar_nb.txt` as the source of data. This file contains comma separated values in each line:

```
# Input file containing data
input_file = 'data_multivar_nb.txt'
```

Let's load the data from this file:

```
# Load data from input file
data = np.loadtxt(input_file, delimiter=',')
X, y = data[:, :-1], data[:, -1]
```

Create an instance of the Naïve Bayes classifier. We will be using the Gaussian Naïve Bayes classifier here. In this type of classifier, we assume that the values associated in each class follow a Gaussian distribution:

```
# Create Naïve Bayes classifier
classifier = GaussianNB()
```

Train the classifier using the training data:

```
# Train the classifier
classifier.fit(X, y)
```

Run the classifier on the training data and predict the output:

```
# Predict the values for training data
y_pred = classifier.predict(X)
```

Let's compute the accuracy of the classifier by comparing the predicted values with the true labels, and then visualize the performance:

```
# Compute accuracy
accuracy = 100.0 * (y == y_pred).sum() / X.shape[0]
print("Accuracy of Naïve Bayes classifier =", round(accuracy, 2), "%")

# Visualize the performance of the classifier
visualize_classifier(classifier, X, y)
```

The preceding method to compute the accuracy of the classifier is not very robust. We need to perform cross validation, so that we don't use the same training data when we are testing it.

Split the data into training and testing subsets. As specified by the `test_size` parameter in the line below, we will allocate 80% for training and the remaining 20% for testing. We'll then train a Naïve Bayes classifier on this data:

```
# Split data into training and test data
X_train, X_test, y_train, y_test = cross_validation.train_test_split(X, y,
test_size=0.2, random_state=3)
classifier_new = GaussianNB()
classifier_new.fit(X_train, y_train)
y_test_pred = classifier_new.predict(X_test)
```

Compute the accuracy of the classifier and visualize the performance:

```
# compute accuracy of the classifier
accuracy = 100.0 * (y_test == y_test_pred).sum() / X_test.shape[0]
print("Accuracy of the new classifier =", round(accuracy, 2), "%")

# Visualize the performance of the classifier
visualize_classifier(classifier_new, X_test, y_test)
```

Let's use the inbuilt functions to calculate the accuracy, precision, and recall values based on threefold cross validation:

```
num_folds = 3
accuracy_values = cross_validation.cross_val_score(classifier,
        X, y, scoring='accuracy', cv=num_folds)
print("Accuracy: " + str(round(100*accuracy_values.mean(), 2)) + "%")

precision_values = cross_validation.cross_val_score(classifier,
        X, y, scoring='precision_weighted', cv=num_folds)
print("Precision: " + str(round(100*precision_values.mean(), 2)) + "%")

recall_values = cross_validation.cross_val_score(classifier,
        X, y, scoring='recall_weighted', cv=num_folds)
print("Recall: " + str(round(100*recall_values.mean(), 2)) + "%")

f1_values = cross_validation.cross_val_score(classifier,
        X, y, scoring='f1_weighted', cv=num_folds)
print("F1: " + str(round(100*f1_values.mean(), 2)) + "%")
```

If you run the code, you will see this for the first training run:

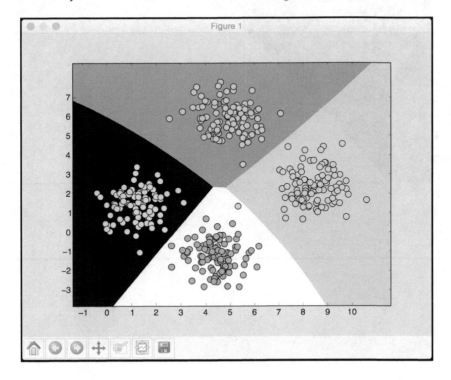

The preceding screenshot shows the boundaries obtained from the classifier. We can see that they separate the 4 clusters well and create regions with boundaries based on the distribution of the input datapoints. You will see in the following screenshot the second training run with cross validation:

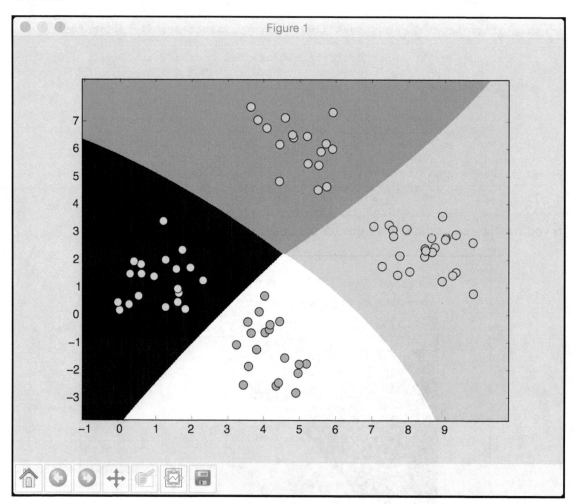

You will see the following printed on your Terminal:

```
Accuracy of Naïve Bayes classifier = 99.75 %
Accuracy of the new classifier = 100.0 %
Accuracy: 99.75%
Precision: 99.76%
Recall: 99.75%
F1: 99.75%
```

The code for this section is given in the file `naive_bayes.py`.

Confusion matrix

A **Confusion matrix** is a figure or a table that is used to describe the performance of a classifier. It is usually extracted from a test dataset for which the ground truth is known. We compare each class with every other class and see how many samples are misclassified. During the construction of this table, we actually come across several key metrics that are very important in the field of machine learning. Let's consider a binary classification case where the output is either *0* or *1*:

- **True positives**: These are the samples for which we predicted *1* as the output and the ground truth is *1* too.
- **True negatives**: These are the samples for which we predicted *0* as the output and the ground truth is *0* too.
- **False positives**: These are the samples for which we predicted *1* as the output but the ground truth is *0*. This is also known as a *Type I error*.
- **False negatives**: These are the samples for which we predicted *0* as the output but the ground truth is *1*. This is also known as a *Type II error*.

Depending on the problem at hand, we may have to optimize our algorithm to reduce the false positive or the false negative rate. For example, in a biometric identification system, it is very important to avoid false positives, because the wrong people might get access to sensitive information. Let's see how to create a confusion matrix.

Create a new Python file and import the following packages:

```
import numpy as np
import matplotlib.pyplot as plt
from sklearn.metrics import confusion_matrix
from sklearn.metrics import classification_report
```

Define some samples labels for the ground truth and the predicted output:

```
# Define sample labels
true_labels = [2, 0, 0, 2, 4, 4, 1, 0, 3, 3, 3]
pred_labels = [2, 1, 0, 2, 4, 3, 1, 0, 1, 3, 3]
```

Create the confusion matrix using the labels we just defined:

```
# Create confusion matrix
confusion_mat = confusion_matrix(true_labels, pred_labels)
```

Visualize the confusion matrix:

```
# Visualize confusion matrix
plt.imshow(confusion_mat, interpolation='nearest', cmap=plt.cm.gray)
plt.title('Confusion matrix')
plt.colorbar()
ticks = np.arange(5)
plt.xticks(ticks, ticks)
plt.yticks(ticks, ticks)
plt.ylabel('True labels')
plt.xlabel('Predicted labels')
plt.show()
```

In the above visualization code, the `ticks` variable refers to the number of distinct classes. In our case, we have five distinct labels.

Let's print the classification report:

```
# Classification report
targets = ['Class-0', 'Class-1', 'Class-2', 'Class-3', 'Class-4']
print('\n', classification_report(true_labels, pred_labels,
target_names=targets))
```

The classification report prints the performance for each class. If you run the code, you will see the following screenshot:

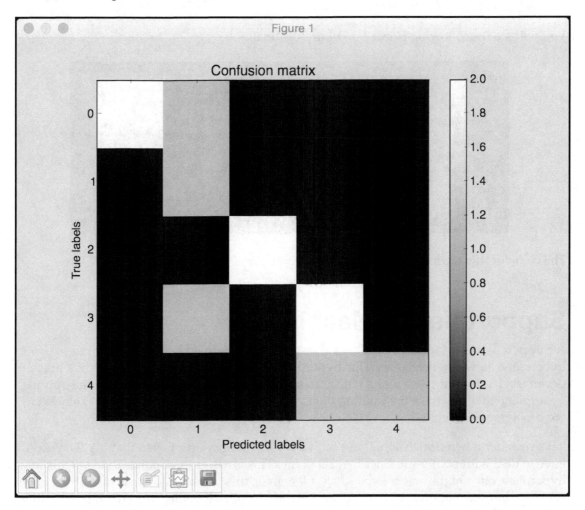

White indicates higher values, whereas black indicates lower values as seen on the color map slider. In an ideal scenario, the diagonal squares will be all white and everything else will be black. This indicates 100% accuracy.

You will see the following printed on your Terminal:

```
             precision    recall  f1-score   support

    Class-0       1.00      0.67      0.80         3
    Class-1       0.33      1.00      0.50         1
    Class-2       1.00      1.00      1.00         2
    Class-3       0.67      0.67      0.67         3
    Class-4       1.00      0.50      0.67         2

avg / total       0.85      0.73      0.75        11
```

The code for this section is given in the file `confusion_matrix.py`.

Support Vector Machines

A **Support Vector Machine (SVM)** is a classifier that is defined using a separating hyperplane between the classes. This **hyperplane** is the N-dimensional version of a line. Given labeled training data and a binary classification problem, the SVM finds the optimal hyperplane that separates the training data into two classes. This can easily be extended to the problem with N classes.

Let's consider a two-dimensional case with two classes of points. Given that it's 2D, we only have to deal with points and lines in a 2D plane. This is easier to visualize than vectors and hyperplanes in a high-dimensional space. Of course, this is a simplified version of the SVM problem, but it is important to understand it and visualize it before we can apply it to high-dimensional data.

Consider the following figure:

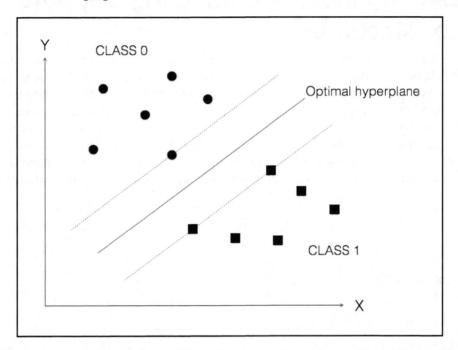

There are two classes of points and we want to find the optimal hyperplane to separate the two classes. But how do we define optimal? In this picture, the solid line represents the best hyperplane. You can draw many different lines to separate the two classes of points, but this line is the best separator, because it maximizes the distance of each point from the separating line. The points on the dotted lines are called Support Vectors. The perpendicular distance between the two dotted lines is called maximum margin.

Classifying income data using Support Vector Machines

We will build a Support Vector Machine classifier to predict the income bracket of a given person based on 14 attributes. Our goal is to see where the income is higher or lower than $50,000 per year. Hence this is a binary classification problem. We will be using the census income dataset available at `https://archive.ics.uci.edu/ml/datasets/Census+Income`. One thing to note in this dataset is that each `datapoint` is a mixture of words and numbers. We cannot use the data in its raw format, because the algorithms don't know how to deal with words. We cannot convert everything using label encoder because numerical data is valuable. Hence we need to use a combination of label encoders and raw numerical data to build an effective classifier.

Create a new Python file and import the following packages:

```
import numpy as np
import matplotlib.pyplot as plt
from sklearn import preprocessing
from sklearn.svm import LinearSVC
from sklearn.multiclass import OneVsOneClassifier
from sklearn import cross_validation
```

We will be using the file `income_data.txt` to load the data. This file contains the income details:

```
# Input file containing data
input_file = 'income_data.txt'
```

In order to load the data from the file, we need to preprocess it so that we can prepare it for classification. We will use at most 25,000 data points for each class:

```
# Read the data
X = []
y = []
count_class1 = 0
count_class2 = 0
max_datapoints = 25000
```

Open the file and start reading the lines:

```
with open(input_file, 'r') as f:
    for line in f.readlines():
        if count_class1 >= max_datapoints and count_class2 >= 
max_datapoints:
            break

        if '?' in line:
            continue
```

Each line is comma separated, so we need to split it accordingly. The last element in each line represents the label. Depending on that label, we will assign it to a class:

```
        data = line[:-1].split(', ')

        if data[-1] == '<=50K' and count_class1 < max_datapoints:
            X.append(data)
            count_class1 += 1

        if data[-1] == '>50K' and count_class2 < max_datapoints:
            X.append(data)
            count_class2 += 1
```

Convert the list into a `numpy` array so that we can give it as an input to the `sklearn` function:

```
# Convert to numpy array
X = np.array(X)
```

If any attribute is a string, then we need to encode it. If it is a number, we can keep it as it is. Note that we will end up with multiple label encoders and we need to keep track of all of them:

```
# Convert string data to numerical data
label_encoder = []
X_encoded = np.empty(X.shape)
for i,item in enumerate(X[0]):
    if item.isdigit():
        X_encoded[:, i] = X[:, i]
    else:
        label_encoder.append(preprocessing.LabelEncoder())
        X_encoded[:, i] = label_encoder[-1].fit_transform(X[:, i])

X = X_encoded[:, :-1].astype(int)
y = X_encoded[:, -1].astype(int)
```

Create the SVM classifier with a linear kernel:

```
# Create SVM classifier
classifier = OneVsOneClassifier(LinearSVC(random_state=0))
```

Train the classifier:

```
# Train the classifier
classifier.fit(X, y)
```

Perform cross validation using an 80/20 split for training and testing, and then predict the output for training data:

```
# Cross validation
X_train, X_test, y_train, y_test = cross_validation.train_test_split(X, y,
test_size=0.2, random_state=5)
classifier = OneVsOneClassifier(LinearSVC(random_state=0))
classifier.fit(X_train, y_train)
y_test_pred = classifier.predict(X_test)
```

Compute the F1 score for the classifier:

```
# Compute the F1 score of the SVM classifier
f1 = cross_validation.cross_val_score(classifier, X, y,
scoring='f1_weighted', cv=3)
print("F1 score: " + str(round(100*f1.mean(), 2)) + "%")
```

Now that the classifier is ready, let's see how to take a random input data point and predict the output. Let's define one such data point:

```
# Predict output for a test datapoint
input_data = ['37', 'Private', '215646', 'HS-grad', '9', 'Never-married',
'Handlers-cleaners', 'Not-in-family', 'White', 'Male', '0', '0', '40',
'United-States']
```

Before we can perform prediction, we need to encode this data point using the label encoders we created earlier:

```
# Encode test datapoint
input_data_encoded = [-1] * len(input_data)
count = 0
for i, item in enumerate(input_data):
    if item.isdigit():
        input_data_encoded[i] = int(input_data[i])
    else:
        input_data_encoded[i] =
int(label_encoder[count].transform(input_data[i]))
        count += 1
```

```
input_data_encoded = np.array(input_data_encoded)
```

We are now ready to predict the output using the classifier:

```
# Run classifier on encoded datapoint and print output
predicted_class = classifier.predict(input_data_encoded)
print(label_encoder[-1].inverse_transform(predicted_class)[0])
```

If you run the code, it will take a few seconds to train the classifier. Once it's done, you will see the following printed on your Terminal:

F1 score: 66.82%

You will also see the output for the test data point:

```
<=50K
```

If you check the values in that data point, you will see that it closely corresponds to the data points in the less than 50K class. You can change the performance of the classifier (F1 score, precision, or recall) by using various different kernels and trying out multiple combinations of the parameters.

The code for this section is given in the file `income_classifier.py`.

What is Regression?

Regression is the process of estimating the relationship between input and output variables. One thing to note is that the output variables are continuous-valued real numbers. Hence there are an infinite number of possibilities. This is in contrast with classification, where the number of output classes is fixed. The classes belong to a finite set of possibilities.

In regression, it is assumed that the output variables depend on the input variables, so we want to see how they are related. Consequently, the input variables are called independent variables, also known as predictors, and output variables are called dependent variables, also known as criterion variables. It is not necessary that the input variables are independent of each other. There are a lot of situations where there are correlations between input variables.

Regression analysis helps us in understanding how the value of the output variable changes when we vary some input variables while keeping other input variables fixed. In linear regression, we assume that the relationship between input and output is linear. This puts a constraint on our modeling procedure, but it's fast and efficient.

Sometimes, linear regression is not sufficient to explain the relationship between input and output. Hence we use polynomial regression, where we use a polynomial to explain the relationship between input and output. This is more computationally complex, but gives higher accuracy. Depending on the problem at hand, we use different forms of regression to extract the relationship. Regression is frequently used for prediction of prices, economics, variations, and so on.

Building a single variable regressor

Let's see how to build a single variable regression model. Create a new Python file and import the following packages:

```
import pickle

import numpy as np
from sklearn import linear_model
import sklearn.metrics as sm
import matplotlib.pyplot as plt
```

We will use the file `data_singlevar_regr.txt` provided to you. This is our source of data:

```
# Input file containing data
input_file = 'data_singlevar_regr.txt'
```

It's a comma-separated file, so we can easily load it using a one-line function call:

```
# Read data
data = np.loadtxt(input_file, delimiter=',')
X, y = data[:, :-1], data[:, -1]
```

Split it into training and testing:

```
# Train and test split
num_training = int(0.8 * len(X))
num_test = len(X) - num_training

# Training data
X_train, y_train = X[:num_training], y[:num_training]

# Test data
X_test, y_test = X[num_training:], y[num_training:]
```

Create a linear regressor object and train it using the training data:

```
# Create linear regressor object
regressor = linear_model.LinearRegression()

# Train the model using the training sets
regressor.fit(X_train, y_train)
```

Predict the output for the testing dataset using the training model:

```
# Predict the output
y_test_pred = regressor.predict(X_test)
```

Plot the output:

```
# Plot outputs
plt.scatter(X_test, y_test, color='green')
plt.plot(X_test, y_test_pred, color='black', linewidth=4)
plt.xticks(())
plt.yticks(())
plt.show()
```

Compute the performance metrics for the regressor by comparing the ground truth, which refers to the actual outputs, with the predicted outputs:

```
# Compute performance metrics
print("Linear regressor performance:")
print("Mean absolute error =", round(sm.mean_absolute_error(y_test,
y_test_pred), 2))
print("Mean squared error =", round(sm.mean_squared_error(y_test,
y_test_pred), 2))
print("Median absolute error =", round(sm.median_absolute_error(y_test,
y_test_pred), 2))
print("Explain variance score =", round(sm.explained_variance_score(y_test,
y_test_pred), 2))
print("R2 score =", round(sm.r2_score(y_test, y_test_pred), 2))
```

Once the model has been created, we can save it into a file so that we can use it later. Python provides a nice module called `pickle` that enables us to do this:

```
# Model persistence
output_model_file = 'model.pkl'

# Save the model
with open(output_model_file, 'wb') as f:
    pickle.dump(regressor, f)
```

Let's load the model from the file on the disk and perform prediction:

```
# Load the model
with open(output_model_file, 'rb') as f:
    regressor_model = pickle.load(f)

# Perform prediction on test data
y_test_pred_new = regressor_model.predict(X_test)
print("\nNew mean absolute error =", round(sm.mean_absolute_error(y_test,
y_test_pred_new), 2))
```

If you run the code, you will see the following screenshot:

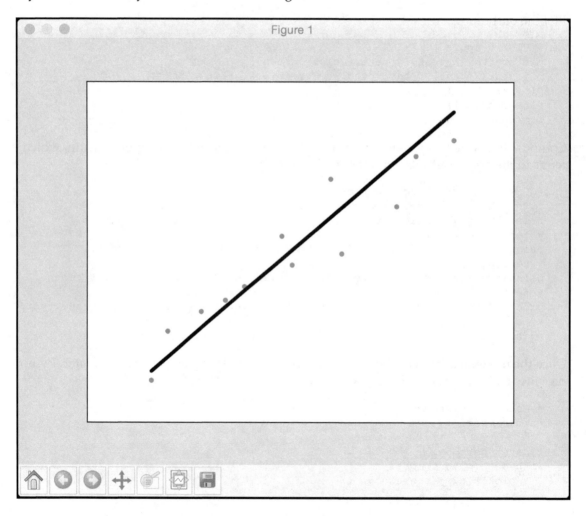

You will see the following printed on your Terminal:

```
Linear regressor performance:
Mean absolute error = 0.59
Mean squared error = 0.49
Median absolute error = 0.51
Explain variance score = 0.86
R2 score = 0.86
New mean absolute error = 0.59
```

The code for this section is given in the file `regressor_singlevar.py`.

Building a multivariable regressor

In the previous section, we discussed how to build a regression model for a single variable. In this section, we will deal with multidimensional data. Create a new Python file and import the following packages:

```
import numpy as np
from sklearn import linear_model
import sklearn.metrics as sm
from sklearn.preprocessing import PolynomialFeatures
```

We will use the file `data_multivar_regr.txt` provided to you.

```
# Input file containing data
input_file = 'data_multivar_regr.txt'
```

This is a comma-separated file, so we can load it easily with a one-line function call:

```
# Load the data from the input file
data = np.loadtxt(input_file, delimiter=',')
X, y = data[:, :-1], data[:, -1]
```

Split the data into training and testing:

```
# Split data into training and testing
num_training = int(0.8 * len(X))
num_test = len(X) - num_training

# Training data
X_train, y_train = X[:num_training], y[:num_training]

# Test data
X_test, y_test = X[num_training:], y[num_training:]
```

[59]

Create and train the linear regressor model:

```
# Create the linear regressor model
linear_regressor = linear_model.LinearRegression()

# Train the model using the training sets
linear_regressor.fit(X_train, y_train)
```

Predict the output for the test dataset:

```
# Predict the output
y_test_pred = linear_regressor.predict(X_test)
```

Print the performance metrics:

```
# Measure performance
print("Linear Regressor performance:")
print("Mean absolute error =", round(sm.mean_absolute_error(y_test,
y_test_pred), 2))
print("Mean squared error =", round(sm.mean_squared_error(y_test,
y_test_pred), 2))
print("Median absolute error =", round(sm.median_absolute_error(y_test,
y_test_pred), 2))
print("Explained variance score =",
round(sm.explained_variance_score(y_test, y_test_pred), 2))
print("R2 score =", round(sm.r2_score(y_test, y_test_pred), 2))
```

Create a polynomial regressor of degree 10. Train the regressor on the training dataset. Let's take a sample data point and see how to perform prediction. The first step is to transform it into a polynomial:

```
# Polynomial regression
polynomial = PolynomialFeatures(degree=10)
X_train_transformed = polynomial.fit_transform(X_train)
datapoint = [[7.75, 6.35, 5.56]]
poly_datapoint = polynomial.fit_transform(datapoint)
```

If you look closely, this data point is very close to the data point on line 11 in our data file, which is [7.66, 6.29, 5.66]. So, a good regressor should predict an output that's close to 41.35. Create a linear regressor object and perform the polynomial fit. Perform the prediction using both linear and polynomial regressors to see the difference:

```
poly_linear_model = linear_model.LinearRegression()
poly_linear_model.fit(X_train_transformed, y_train)
print("\nLinear regression:\n", linear_regressor.predict(datapoint))
print("\nPolynomial regression:\n",
poly_linear_model.predict(poly_datapoint))
```

If you run the code, you will see the following printed on your Terminal:

```
Linear Regressor performance:
Mean absolute error = 3.58
Mean squared error = 20.31
Median absolute error = 2.99
Explained variance score = 0.86
R2 score = 0.86
```

You will see the following as well:

```
Linear regression:
 [ 36.05286276]
Polynomial regression:
 [ 41.46961676]
```

As you can see, the polynomial regressor is closer to 41.35. The code for this section is given in the file `regressor_multivar.py`.

Estimating housing prices using a Support Vector Regressor

Let's see how to use the SVM concept to build a regressor to estimate the housing prices. We will use the dataset available in `sklearn` where each data point is define, by 13 attributes. Our goal is to estimate the housing prices based on these attributes.

Create a new Python file and import the following packages:

```
import numpy as np
from sklearn import datasets
from sklearn.svm import SVR
from sklearn.metrics import mean_squared_error, explained_variance_score
from sklearn.utils import shuffle
```

Load the housing dataset:

```
# Load housing data
data = datasets.load_boston()
```

Let's shuffle the data so that we don't bias our analysis:

```
# Shuffle the data
X, y = shuffle(data.data, data.target, random_state=7)
```

Classification and Regression Using Supervised Learning

Split the dataset into training and testing in an 80/20 format:

```
# Split the data into training and testing datasets
num_training = int(0.8 * len(X))
X_train, y_train = X[:num_training], y[:num_training]
X_test, y_test = X[num_training:], y[num_training:]
```

Create and train the Support Vector Regressor using a linear kernel. The C parameter represents the penalty for training error. If you increase the value of C, the model will fine-tune it more to fit the training data. But this might lead to overfitting and cause it to lose its generality. The epsilon parameter specifies a threshold; there is no penalty for training error if the predicted value is within this distance from the actual value:

```
# Create Support Vector Regression model
sv_regressor = SVR(kernel='linear', C=1.0, epsilon=0.1)

# Train Support Vector Regressor
sv_regressor.fit(X_train, y_train)
```

Evaluate the performance of the regressor and print the metrics:

```
# Evaluate performance of Support Vector Regressor
y_test_pred = sv_regressor.predict(X_test)
mse = mean_squared_error(y_test, y_test_pred)
evs = explained_variance_score(y_test, y_test_pred)
print("\n#### Performance ####")
print("Mean squared error =", round(mse, 2))
print("Explained variance score =", round(evs, 2))
```

Let's take a test data point and perform prediction:

```
# Test the regressor on test datapoint
test_data = [3.7, 0, 18.4, 1, 0.87, 5.95, 91, 2.5052, 26, 666, 20.2, 351.34, 15.27]
print("\nPredicted price:", sv_regressor.predict([test_data])[0])
```

If you run the code, you will see the following printed on the Terminal:

```
#### Performance ####
Mean squared error = 15.41
Explained variance score = 0.82
Predicted price: 18.5217801073
```

The code for this section is given in the file `house_prices.py`.

[62]

Summary

In this chapter, we learned the difference between supervised and unsupervised learning. We discussed the data classification problem and how to solve it. We understood how to preprocess data using various methods. We also learned about label encoding and how to build a label encoder. We discussed logistic regression and built a logistic regression classifier. We understood what Naïve Bayes classifier is and learned how to build it. We also learned how to build a confusion matrix.

We discussed Support Vector Machines and understood how to build a classifier based on that. We learned about regression and understood how to use linear and polynomial regression for single and multivariable data. We then used Support Vector Regressor to estimate the housing prices using input attributes.

In the next chapter, we will learn about predictive analytics and how to build a predictive engine using ensemble learning.

3
Predictive Analytics with Ensemble Learning

In this chapter, we are going to learn about Ensemble Learning and how to use it for predictive analytics. By the end of this chapter, you will know these topics:

- Building learning models with Ensemble Learning
- What are Decision Trees and how to build a Decision Trees classifier
- What are Random Forests and Extremely Random Forests, and how to build classifiers based on them
- Estimating the confidence measure of the predictions
- Dealing with class imbalance
- Finding optimal training parameters using grid search
- Computing relative feature importance
- Predicting traffic using Extremely Random Forests regressor

What is Ensemble Learning?

Ensemble Learning refers to the process of building multiple models and then combining them in a way that can produce better results than individual models. These individual models can be classifiers, regressors, or anything else that models data in some way. Ensemble learning is used extensively across multiple fields including data classification, predictive modeling, anomaly detection, and so on.

Why do we need ensemble learning in the first place? In order to understand this, let's take a real-life example. You want to buy a new TV, but you don't know what the latest models are. Your goal is to get the best value for your money, but you don't have enough knowledge on this topic to make an informed decision. When you have to make a decision about something like this, you go around and try to get the opinions of multiple experts in the domain. This will help you make the best decision. More often than not, instead of just relying on a single opinion, you tend to make a final decision by combining the individual decisions of those experts. The reason we do that is because we want to minimize the possibility of a wrong or suboptimal decision.

Building learning models with Ensemble Learning

When we select a model, the most commonly used procedure is to choose the one with the smallest error on the training dataset. The problem with this approach is that it will not always work. The model might get biased or overfit the training data. Even when we compute the model using cross validation, it can perform poorly on unknown data.

One of the main reasons ensemble learning is so effective is because it reduces the overall risk of making a poor model selection. This enables it to train in a diverse manner and then perform well on unknown data. When we build a model using ensemble learning, the individual models need to exhibit some diversity. This would allow them to capture various nuances in our data; hence the overall model becomes more accurate.

The diversity is achieved by using different training parameters for each individual model. This allows individual models to generate different decision boundaries for training data. This means that each model will use different rules to make an inference, which is a powerful way of validating the final result. If there is agreement among the models, then we know that the output is correct.

What are Decision Trees?

A **Decision Tree** is a structure that allows us to split the dataset into branches and then make simple decisions at each level. This will allow us to arrive at the final decision by walking down the tree. Decision Trees are produced by training algorithms, which identify how we can split the data in the best possible way.

Any decision process starts at the root node at the top of the tree. Each node in the tree is basically a decision rule. Algorithms construct these rules based on the relationship between the input data and the target labels in the training data. The values in the input data are utilized to estimate the value for the output.

Now that we understand basic concept of Decision Trees, the next thing is to understand how the trees are automatically constructed. We need algorithms that can construct the optimal tree based on our data. In order to understand it, we need to understand the concept of entropy. In this context, entropy refers to information entropy and not thermodynamic entropy. Entropy is basically a measure of uncertainty. One of the main goals of a decision tree is to reduce uncertainty as we move from the root node towards the leaf nodes. When we see an unknown data point, we are completely uncertain about the output. By the time we reach the leaf node, we are certain about the output. This means that we need to construct the decision tree in a way that will reduce the uncertainty at each level. This implies that we need to reduce the entropy as we progress down the tree.

You can learn more about this at: `https://prateekvjoshi.com/2016/03/22/how-are-decision-trees-constructed-in-machine-learning`.

Building a Decision Tree classifier

Let's see how to build a classifier using Decision Trees in Python. Create a new Python file and import the following packages:

```
import numpy as np
import matplotlib.pyplot as plt
from sklearn.metrics import classification_report
from sklearn import cross_validation
from sklearn.tree import DecisionTreeClassifier

from utilities import visualize_classifier
```

We will be using the data in the `data_decision_trees.txt` file that's provided to you. In this file, each line contains comma-separated values. The first two values correspond to the input data and the last value corresponds to the target label. Let's load the data from that file:

```
# Load input data
input_file = 'data_decision_trees.txt'
data = np.loadtxt(input_file, delimiter=',')
X, y = data[:, :-1], data[:, -1]
```

Separate the input data into two separate classes based on the labels:

```
# Separate input data into two classes based on labels
class_0 = np.array(X[y==0])
class_1 = np.array(X[y==1])
```

Let's visualize the input data using a scatter plot:

```
# Visualize input data
plt.figure()
plt.scatter(class_0[:, 0], class_0[:, 1], s=75, facecolors='black',
        edgecolors='black', linewidth=1, marker='x')
plt.scatter(class_1[:, 0], class_1[:, 1], s=75, facecolors='white',
        edgecolors='black', linewidth=1, marker='o')
plt.title('Input data')
```

We need to split the data into training and testing datasets:

```
# Split data into training and testing datasets
X_train, X_test, y_train, y_test = cross_validation.train_test_split(
        X, y, test_size=0.25, random_state=5)
```

Create, build, and visualize a decision tree classifier based on the training dataset. The `random_state` parameter refers to the seed used by the random number generator required for the initialization of the decision tree classification algorithm. The `max_depth` parameter refers to the maximum depth of the tree that we want to construct:

```
# Decision Trees classifier
params = {'random_state': 0, 'max_depth': 4}
classifier = DecisionTreeClassifier(**params)
classifier.fit(X_train, y_train)
visualize_classifier(classifier, X_train, y_train, 'Training dataset')
```

Compute the output of the classifier on the test dataset and visualize it:

```
y_test_pred = classifier.predict(X_test)
visualize_classifier(classifier, X_test, y_test, 'Test dataset')
```

Evaluate the performance of the classifier by printing the classification report:

```
# Evaluate classifier performance
class_names = ['Class-0', 'Class-1']
print("\n" + "#"*40)
print("\nClassifier performance on training dataset\n")
print(classification_report(y_train, classifier.predict(X_train),
    target_names=class_names))
print("#"*40 + "\n")
```

```
print("#"*40)
print("\nClassifier performance on test dataset\n")
print(classification_report(y_test, y_test_pred, target_names=class_names))
print("#"*40 + "\n")

plt.show()
```

The full code is given in the `decision_trees.py` file. If you run the code, you will see a few figures. The first screenshot is the visualization of input data:

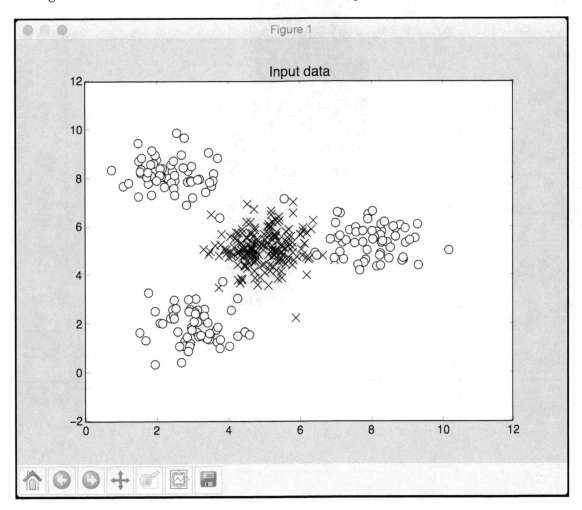

The second screenshot shows the classifier boundaries on the test dataset:

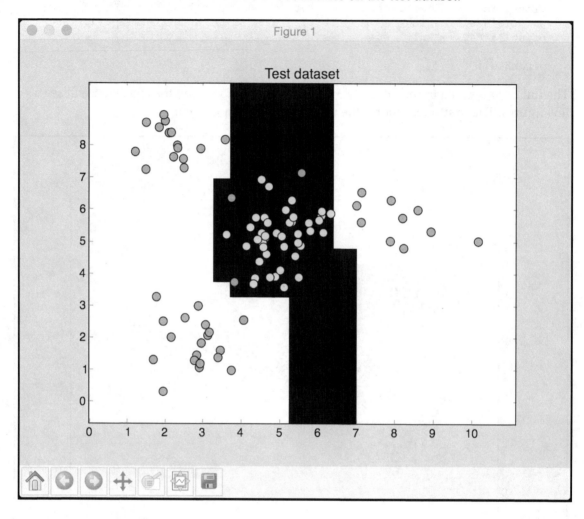

You will see the following printed on your Terminal:

```
##########################################

Classifier performance on training dataset

             precision    recall  f1-score   support

    Class-0       0.99      1.00      1.00       137
    Class-1       1.00      0.99      1.00       133

avg / total       1.00      1.00      1.00       270

##########################################

##########################################

Classifier performance on test dataset

             precision    recall  f1-score   support

    Class-0       0.93      1.00      0.97        43
    Class-1       1.00      0.94      0.97        47

avg / total       0.97      0.97      0.97        90

##########################################
```

The performance of a classifier is characterized by `precision`, `recall`, and `f1-scores`. Precision refers to the accuracy of the classification and recall refers to the number of items that were retrieved as a percentage of the overall number of items that were supposed to be retrieved. A good classifier will have high precision and high recall, but it is usually a trade-off between the two. Hence we have `f1-score` to characterize that. F1 score is the harmonic mean of precision and recall, which gives it a good balance between precision and recall values.

What are Random Forests and Extremely Random Forests?

A **Random Forest** is a particular instance of ensemble learning where individual models are constructed using Decision Trees. This ensemble of Decision Trees is then used to predict the output value. We use a random subset of training data to construct each Decision Tree. This will ensure diversity among various decision trees. In the first section, we discussed that one of the most important things in ensemble learning is to ensure that there's diversity among individual models.

One of the best things about Random Forests is that they do not overfit. As we know, overfitting is a problem that we encounter frequently in machine learning. By constructing a diverse set of Decision Trees using various random subsets, we ensure that the model does not overfit the training data. During the construction of the tree, the nodes are split successively and the best thresholds are chosen to reduce the entropy at each level. This split doesn't consider all the features in the input dataset. Instead, it chooses the best split among the random subset of the features that is under consideration. Adding this randomness tends to increase the bias of the random forest, but the variance decreases because of averaging. Hence, we end up with a robust model.

Extremely Random Forests take randomness to the next level. Along with taking a random subset of features, the thresholds are chosen at random too. These randomly generated thresholds are chosen as the splitting rules, which reduce the variance of the model even further. Hence the decision boundaries obtained using Extremely Random Forests tend to be smoother than the ones obtained using Random Forests.

Building Random Forest and Extremely Random Forest classifiers

Let's see how to build a classifier based on Random Forests and Extremely Random Forests. The way to construct both classifiers is very similar, so we will use an input flag to specify which classifier needs to be built.

Create a new Python file and import the following packages:

```
import argparse

import numpy as np
import matplotlib.pyplot as plt
from sklearn.metrics import classification_report
from sklearn import cross_validation
from sklearn.ensemble import RandomForestClassifier, ExtraTreesClassifier
from sklearn import cross_validation
from sklearn.metrics import classification_report

from utilities import visualize_classifier
```

Define an argument parser for Python so that we can take the classifier type as an input parameter. Depending on this parameter, we can construct a Random Forest classifier or an Extremely Random forest classifier:

```
# Argument parser
def build_arg_parser():
    parser = argparse.ArgumentParser(description='Classify data using \
            Ensemble Learning techniques')
    parser.add_argument('--classifier-type', dest='classifier_type',
            required=True, choices=['rf', 'erf'], help="Type of classifier
                    \to use; can be either 'rf' or 'erf'")
    return parser
```

Define the main function and parse the input arguments:

```
if __name__=='__main__':
    # Parse the input arguments
    args = build_arg_parser().parse_args()
    classifier_type = args.classifier_type
```

We will be using the data from the `data_random_forests.txt` file that is provided to you. Each line in this file contains comma-separated values. The first two values correspond to the input data and the last value corresponds to the target label. We have three distinct classes in this dataset. Let's load the data from that file:

```
    # Load input data
    input_file = 'data_random_forests.txt'
    data = np.loadtxt(input_file, delimiter=',')
    X, y = data[:, :-1], data[:, -1]
```

Separate the input data into three classes:

```
# Separate input data into three classes based on labels
class_0 = np.array(X[y==0])
class_1 = np.array(X[y==1])
class_2 = np.array(X[y==2])
```

Let's visualize the input data:

```
# Visualize input data
plt.figure()
plt.scatter(class_0[:, 0], class_0[:, 1], s=75, facecolors='white',
            edgecolors='black', linewidth=1, marker='s')
plt.scatter(class_1[:, 0], class_1[:, 1], s=75, facecolors='white',
            edgecolors='black', linewidth=1, marker='o')
plt.scatter(class_2[:, 0], class_2[:, 1], s=75, facecolors='white',
            edgecolors='black', linewidth=1, marker='^')
plt.title('Input data')
```

Split the data into training and testing datasets:

```
# Split data into training and testing datasets
X_train, X_test, y_train, y_test = cross_validation.train_test_split(
        X, y, test_size=0.25, random_state=5)
```

Define the parameters to be used when we construct the classifier. The n_estimators parameter refers to the number of trees that will be constructed. The max_depth parameter refers to the maximum number of levels in each tree. The random_state parameter refers to the seed value of the random number generator needed to initialize the random forest classifier algorithm:

```
# Ensemble Learning classifier
params = {'n_estimators': 100, 'max_depth': 4, 'random_state': 0}
```

Depending on the input parameter, we either construct a random forest classifier or an extremely random forest classifier:

```
if classifier_type == 'rf':
    classifier = RandomForestClassifier(**params)
else:
    classifier = ExtraTreesClassifier(**params)
```

Train and visualize the classifier:

```
classifier.fit(X_train, y_train)
visualize_classifier(classifier, X_train, y_train, 'Training
dataset')
```

Compute the output based on the test dataset and visualize it:

```
y_test_pred = classifier.predict(X_test)
visualize_classifier(classifier, X_test, y_test, 'Test dataset')
```

Evaluate the performance of the classifier by printing the classification report:

```
# Evaluate classifier performance
class_names = ['Class-0', 'Class-1', 'Class-2']
print("\n" + "#"*40)
print("\nClassifier performance on training dataset\n")
print(classification_report(y_train, classifier.predict(X_train),
target_names=class_names))
print("#"*40 + "\n")

print("#"*40)
print("\nClassifier performance on test dataset\n")
print(classification_report(y_test, y_test_pred,
target_names=class_names))
print("#"*40 + "\n")
```

The full code is given in the `random_forests.py` file. Let's run the code with the Random Forest classifier using the `rf` flag in the input argument. Run the following command on your Terminal:

```
$ python3 random_forests.py --classifier-type rf
```

You will see a few figures pop up. The first screenshot is the input data:

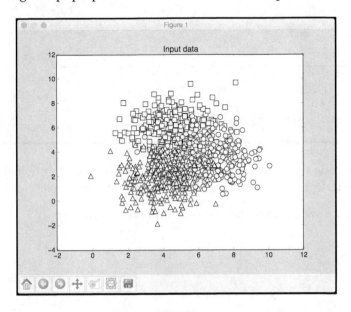

In the preceding screenshot, the three classes are being represented by squares, circles, and triangles. We see that there is a lot of overlap between classes, but that should be fine for now. The second screenshot shows the classifier boundaries:

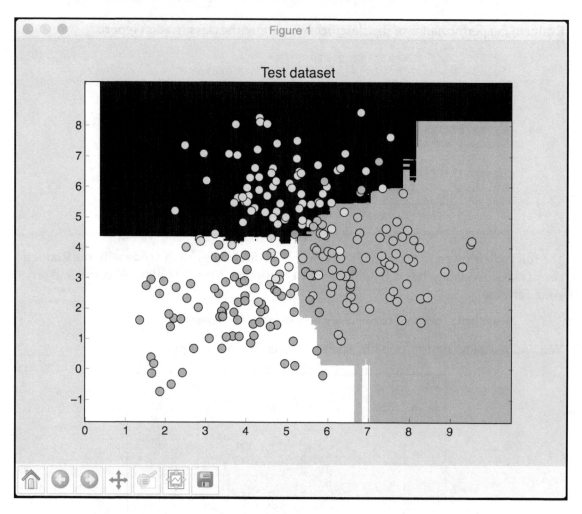

Now let's run the code with the Extremely Random Forest classifier by using the `erf` flag in the input argument. Run the following command on your Terminal:

```
$ python3 random_forests.py --classifier-type erf
```

You will see a few figures pop up. We already know what the input data looks like. The second screenshot shows the classifier boundaries:

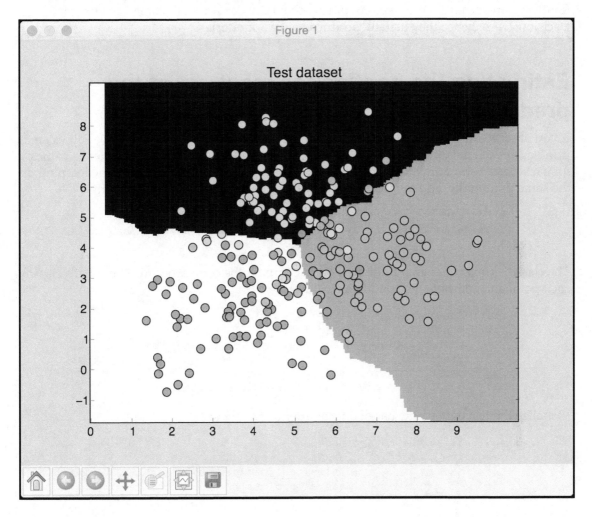

If you compare the preceding screenshot with the boundaries obtained from Random Forest classifier, you will see that these boundaries are smoother. The reason is that Extremely Random Forests have more freedom during the training process to come up with good Decision Trees, hence they usually produce better boundaries.

Estimating the confidence measure of the predictions

If you observe the outputs obtained on the terminal, you will see that the probabilities are printed for each data point. These probabilities are used to measure the confidence values for each class. Estimating the confidence values is an important task in machine learning. In the same python file, add the following line to define an array of test data points:

```
# Compute confidence
test_datapoints = np.array([[5, 5], [3, 6], [6, 4], [7, 2], [4, 4], [5, 2]])
```

The classifier object has an inbuilt method to compute the confidence measure. Let's classify each point and compute the confidence values:

```
print("\nConfidence measure:")
for datapoint in test_datapoints:
    probabilities = classifier.predict_proba([datapoint])[0]
    predicted_class = 'Class-' + str(np.argmax(probabilities))
    print('\nDatapoint:', datapoint)
    print('Predicted class:', predicted_class)
```

Visualize the test data points based on classifier boundaries:

```
# Visualize the datapoints
visualize_classifier(classifier, test_datapoints,
            [0]*len(test_datapoints),
            'Test datapoints')
plt.show()
```

If you run the code with the `rf` flag, you will get the following output:

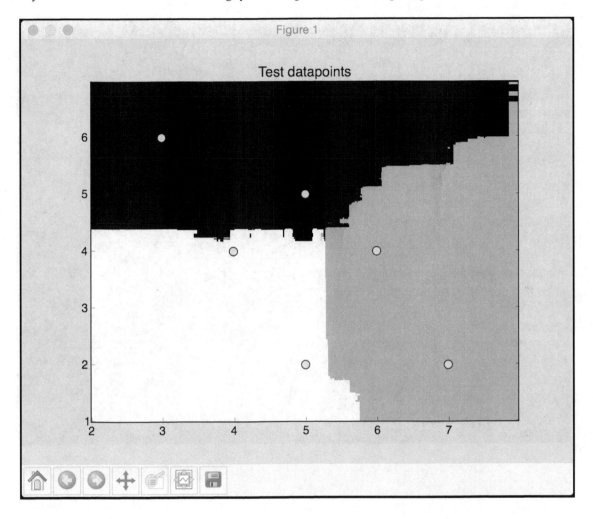

You will get the following output on your Terminal:

```
Datapoint: [5 5]
Probabilities: [ 0.81427532  0.08639273  0.09933195]
Predicted class: Class-0

Datapoint: [3 6]
Probabilities: [ 0.93574458  0.02465345  0.03960197]
Predicted class: Class-0

Datapoint: [6 4]
Probabilities: [ 0.12232404  0.7451078   0.13256816]
Predicted class: Class-1

Datapoint: [7 2]
Probabilities: [ 0.05415465  0.70660226  0.23924309]
Predicted class: Class-1

Datapoint: [4 4]
Probabilities: [ 0.20594744  0.15523491  0.63881765]
Predicted class: Class-2

Datapoint: [5 2]
Probabilities: [ 0.05403583  0.0931115   0.85285267]
Predicted class: Class-2
```

For each data point, it computes the probability of that point belonging to our three classes. We pick the one with the highest confidence. If you run the code with the `erf` flag, you will get the following output:

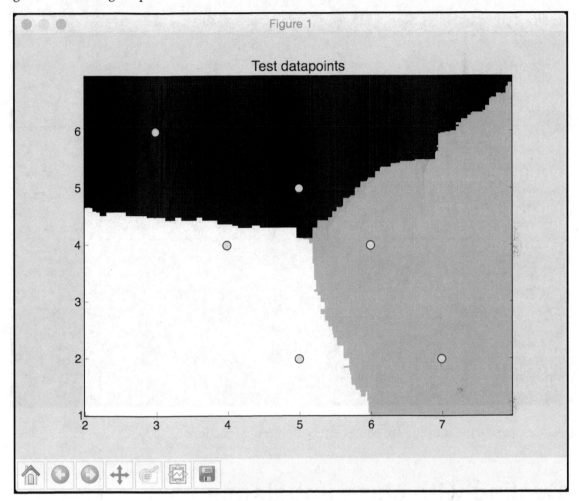

You will get the following output on your Terminal:

```
Datapoint: [5 5]
Probabilities: [ 0.48904419  0.28020114  0.23075467]
Predicted class: Class-0

Datapoint: [3 6]
Probabilities: [ 0.66707383  0.12424406  0.20868211]
Predicted class: Class-0

Datapoint: [6 4]
Probabilities: [ 0.25788769  0.49535144  0.24676087]
Predicted class: Class-1

Datapoint: [7 2]
Probabilities: [ 0.10794013  0.6246677   0.26739217]
Predicted class: Class-1

Datapoint: [4 4]
Probabilities: [ 0.33383778  0.21495182  0.45121039]
Predicted class: Class-2

Datapoint: [5 2]
Probabilities: [ 0.18671115  0.28760896  0.52567989]
Predicted class: Class-2
```

As we can see, the outputs are consistent with our observations.

Dealing with class imbalance

A classifier is only as good as the data that's used for training. One of the most common problems we face in the real world is the quality of data. For a classifier to perform well, it needs to see equal number of points for each class. But when we collect data in the real world, it's not always possible to ensure that each class has the exact same number of data points. If one class has 10 times the number of data points of the other class, then the classifier tends to get biased towards the first class. Hence we need to make sure that we account for this imbalance algorithmically. Let's see how to do that.

Create a new Python file and import the following packages:

```
import sys

import numpy as np
import matplotlib.pyplot as plt
from sklearn.ensemble import ExtraTreesClassifier
from sklearn import cross_validation
from sklearn.metrics import classification_report

from utilities import visualize_classifier
```

We will use the data in the file `data_imbalance.txt` for our analysis. Let's load the data. Each line in this file contains comma-separated values. The first two values correspond to the input data and the last value corresponds to the target label. We have two classes in this dataset. Let's load the data from that file:

```
# Load input data
input_file = 'data_imbalance.txt'
data = np.loadtxt(input_file, delimiter=',')
X, y = data[:, :-1], data[:, -1]
```

Separate the input data into two classes:

```
# Separate input data into two classes based on labels
class_0 = np.array(X[y==0])
class_1 = np.array(X[y==1])
```

Visualize the input data using scatter plot:

```
# Visualize input data
plt.figure()
plt.scatter(class_0[:, 0], class_0[:, 1], s=75, facecolors='black',
            edgecolors='black', linewidth=1, marker='x')
plt.scatter(class_1[:, 0], class_1[:, 1], s=75, facecolors='white',
            edgecolors='black', linewidth=1, marker='o')
plt.title('Input data')
```

Split the data into training and testing datasets:

```
# Split data into training and testing datasets
X_train, X_test, y_train, y_test = cross_validation.train_test_split(
        X, y, test_size=0.25, random_state=5)
```

Next, we define the parameters for the Extremely Random Forest classifier. Note that there is an input parameter called `balance` that controls whether or not we want to algorithmically account for class imbalance. If so, then we need to add another parameter called `class_weight` that tells the classifier that it should balance the weight, so that it's proportional to the number of data points in each class:

```
# Extremely Random Forests classifier
params = {'n_estimators': 100, 'max_depth': 4, 'random_state': 0}
if len(sys.argv) > 1:
    if sys.argv[1] == 'balance':
        params = {'n_estimators': 100, 'max_depth': 4, 'random_state': 0, 'class_weight': 'balanced'}
    else:
        raise TypeError("Invalid input argument; should be 'balance'")
```

Build, train, and visualize the classifier using training data:

```
classifier = ExtraTreesClassifier(**params)
classifier.fit(X_train, y_train)
visualize_classifier(classifier, X_train, y_train, 'Training dataset')
```

Predict the output for test dataset and visualize the output:

```
y_test_pred = classifier.predict(X_test)
visualize_classifier(classifier, X_test, y_test, 'Test dataset')
```

Compute the performance of the classifier and print the classification report:

```
# Evaluate classifier performance
class_names = ['Class-0', 'Class-1']
print("\n" + "#"*40)
print("\nClassifier performance on training dataset\n")
print(classification_report(y_train, classifier.predict(X_train), target_names=class_names))
print("#"*40 + "\n")

print("#"*40)
print("\nClassifier performance on test dataset\n")
print(classification_report(y_test, y_test_pred, target_names=class_names))
print("#"*40 + "\n")

plt.show()
```

The full code is given in the file `class_imbalance.py`. If you run the code, you will see a few screenshots. The first screenshot shows the input data:

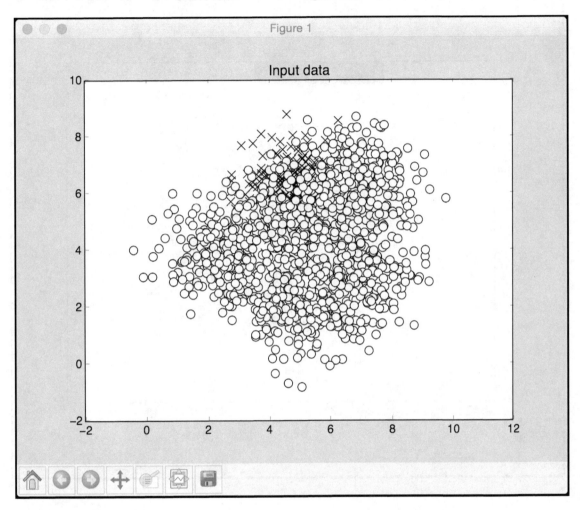

The second screenshot shows the classifier boundary for the test dataset:

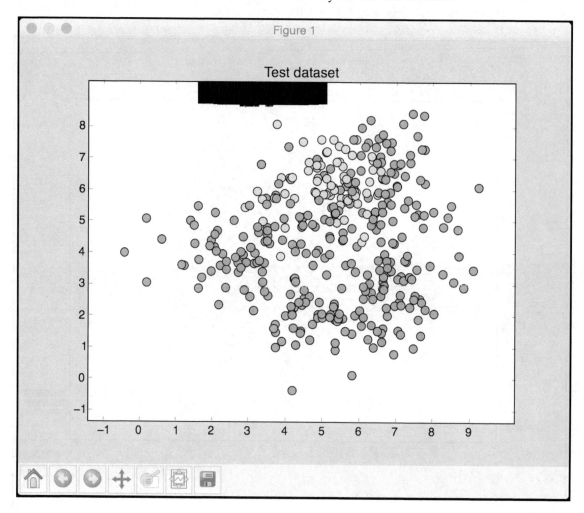

The preceding screenshot indicates that the boundary was not able to capture the actual boundary between the two classes. The black patch near the top represents the boundary. You should see the following output on your Terminal:

```
##########################################

Classifier performance on test dataset

             precision    recall  f1-score   support

    Class-0       0.00      0.00      0.00        69
    Class-1       0.82      1.00      0.90       306

avg / total       0.67      0.82      0.73       375

##########################################
```

You see a warning because the values are *0* in the first row, which leads to a divide-by-zero error (`ZeroDivisionError` exception) when we compute the `f1-score`. Run the code on the terminal using the ignore flag so that you do not see the divide-by-zero warning:

```
$ python3 --W ignore class_imbalance.py
```

Now if you want to account for class imbalance, run it with the balance flag:

```
$ python3 class_imbalance.py balance
```

The classifier output looks like this:

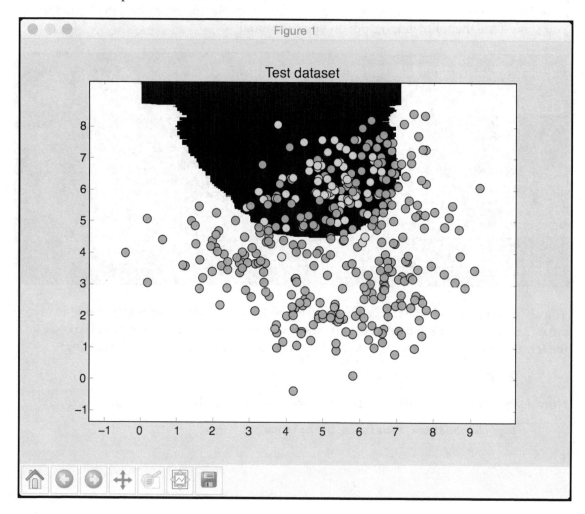

You should see the following output on your Terminal:

```
##########################################
Classifier performance on test dataset
             precision    recall  f1-score   support

    Class-0       0.45      0.94      0.61        69
    Class-1       0.98      0.74      0.84       306

avg / total       0.88      0.78      0.80       375
##########################################
```

By accounting for the class imbalance, we were able to classify the data points in `class-0` with non-zero accuracy.

Finding optimal training parameters using grid search

When you are working with classifiers, you do not always know what the best parameters are. You cannot brute-force it by checking for all possible combinations manually. This is where grid search becomes useful. Grid search allows us to specify a range of values and the classifier will automatically run various configurations to figure out the best combination of parameters. Let's see how to do it.

Create a new Python file and import the following packages:

```python
import numpy as np
import matplotlib.pyplot as plt
from sklearn.metrics import classification_report
from sklearn import cross_validation, grid_search
from sklearn.ensemble import ExtraTreesClassifier
from sklearn import cross_validation
from sklearn.metrics import classification_report

from utilities import visualize_classifier
```

We will use the data available in `data_random_forests.txt` for analysis:

```python
# Load input data
input_file = 'data_random_forests.txt'
data = np.loadtxt(input_file, delimiter=',')
X, y = data[:, :-1], data[:, -1]
```

Separate the data into three classes:

```
# Separate input data into three classes based on labels
class_0 = np.array(X[y==0])
class_1 = np.array(X[y==1])
class_2 = np.array(X[y==2])
```

Split the data into training and testing datasets:

```
# Split the data into training and testing datasets
X_train, X_test, y_train, y_test = cross_validation.train_test_split(
        X, y, test_size=0.25, random_state=5)
```

Specify the grid of parameters that you want the classifier to test. Usually we keep one parameter constant and vary the other parameter. We then do it vice versa to figure out the best combination. In this case, we want to find the best values for n_estimators and max_depth. Let's specify the parameter grid:

```
# Define the parameter grid
parameter_grid = [ {'n_estimators': [100], 'max_depth': [2, 4, 7, 12, 16]},
                   {'max_depth': [4], 'n_estimators': [25, 50, 100, 250]}
                 ]
```

Let's define the metrics that the classifier should use to find the best combination of parameters:

```
metrics = ['precision_weighted', 'recall_weighted']
```

For each metric, we need to run the grid search, where we train the classifier for a particular combination of parameters:

```
for metric in metrics:
    print("\n##### Searching optimal parameters for", metric)

    classifier = grid_search.GridSearchCV(
            ExtraTreesClassifier(random_state=0),
            parameter_grid, cv=5, scoring=metric)
    classifier.fit(X_train, y_train)
```

Print the score for each parameter combination:

```
print("\nGrid scores for the parameter grid:")
for params, avg_score, _ in classifier.grid_scores_:
    print(params, '-->', round(avg_score, 3))

print("\nBest parameters:", classifier.best_params_)
```

Print the performance report:

```
y_pred = classifier.predict(X_test)
print("\nPerformance report:\n")
print(classification_report(y_test, y_pred))
```

The full code is given in the file `run_grid_search.py`. If you run the code, you will get this output on the Terminal for the precision metric:

```
##### Searching optimal parameters for precision_weighted

Grid scores for the parameter grid:
{'n_estimators': 100, 'max_depth': 2} --> 0.847
{'n_estimators': 100, 'max_depth': 4} --> 0.841
{'n_estimators': 100, 'max_depth': 7} --> 0.844
{'n_estimators': 100, 'max_depth': 12} --> 0.836
{'n_estimators': 100, 'max_depth': 16} --> 0.818
{'n_estimators': 25, 'max_depth': 4} --> 0.846
{'n_estimators': 50, 'max_depth': 4} --> 0.84
{'n_estimators': 100, 'max_depth': 4} --> 0.841
{'n_estimators': 250, 'max_depth': 4} --> 0.845

Best parameters: {'n_estimators': 100, 'max_depth': 2}

Performance report:

             precision    recall  f1-score   support

        0.0       0.94      0.81      0.87        79
        1.0       0.81      0.86      0.83        70
        2.0       0.83      0.91      0.87        76

avg / total       0.86      0.86      0.86       225
```

Based on the combinations in the grid search, it will print out the best combination for the precision metric. If we want to know the best combination for recall, we need to check the following output on the Terminal:

```
##### Searching optimal parameters for recall_weighted

Grid scores for the parameter grid:
{'n_estimators': 100, 'max_depth': 2} --> 0.84
{'n_estimators': 100, 'max_depth': 4} --> 0.837
{'n_estimators': 100, 'max_depth': 7} --> 0.841
{'n_estimators': 100, 'max_depth': 12} --> 0.834
{'n_estimators': 100, 'max_depth': 16} --> 0.816
{'n_estimators': 25, 'max_depth': 4} --> 0.843
{'n_estimators': 50, 'max_depth': 4} --> 0.836
{'n_estimators': 100, 'max_depth': 4} --> 0.837
{'n_estimators': 250, 'max_depth': 4} --> 0.841

Best parameters: {'n_estimators': 25, 'max_depth': 4}

Performance report:

             precision    recall  f1-score   support

        0.0       0.93      0.84      0.88        79
        1.0       0.85      0.86      0.85        70
        2.0       0.84      0.92      0.88        76

avg / total       0.87      0.87      0.87       225
```

It is a different combination for recall, which makes sense because precision and recall are different metrics that demand different parameter combinations.

Computing relative feature importance

When we are working with a dataset that contains N-dimensional data points, we have to understand that not all features are equally important. Some are more discriminative than others. If we have this information, we can use it to reduce the dimensional. This is very useful in reducing the complexity and increasing the speed of the algorithm. Sometimes, a few features are completely redundant. Hence they can be easily removed from the dataset.

We will be using the `AdaBoost` regressor to compute feature importance. AdaBoost, short for Adaptive Boosting, is an algorithm that's frequently used in conjunction with other machine learning algorithms to improve their performance. In AdaBoost, the training data points are drawn from a distribution to train the current classifier. This distribution is updated iteratively so that the subsequent classifiers get to focus on the more difficult data points. The difficult data points are the ones that are misclassified. This is done by updating the distribution at each step. This will make the data points that were previously misclassified more likely to come up in the next sample dataset that's used for training. These classifiers are then cascaded and the decision is taken through weighted majority voting.

Create a new Python file and import the following packages:

```
import numpy as np
import matplotlib.pyplot as plt
from sklearn.tree import DecisionTreeRegressor
from sklearn.ensemble import AdaBoostRegressor
from sklearn import datasets
from sklearn.metrics import mean_squared_error, explained_variance_score
from sklearn import cross_validation
from sklearn.utils import shuffle
```

We will use the inbuilt housing dataset available in scikit-learn:

```
# Load housing data
housing_data = datasets.load_boston()
```

Shuffle the data so that we don't bias our analysis:

```
# Shuffle the data
X, y = shuffle(housing_data.data, housing_data.target, random_state=7)
```

Split the dataset into training and testing:

```
# Split data into training and testing datasets
X_train, X_test, y_train, y_test = cross_validation.train_test_split(
        X, y, test_size=0.2, random_state=7)
```

Define and train an `AdaBoostregressor` using the Decision Tree regressor as the individual model:

```
# AdaBoost Regressor model
regressor = AdaBoostRegressor(DecisionTreeRegressor(max_depth=4),
        n_estimators=400, random_state=7)
regressor.fit(X_train, y_train)
```

Estimate the performance of the regressor:

```
# Evaluate performance of AdaBoost regressor
y_pred = regressor.predict(X_test)
mse = mean_squared_error(y_test, y_pred)
evs = explained_variance_score(y_test, y_pred )
print("\nADABOOST REGRESSOR")
print("Mean squared error =", round(mse, 2))
print("Explained variance score =", round(evs, 2))
```

This regressor has an inbuilt method that can be called to compute the relative feature importance:

```
# Extract feature importances
feature_importances = regressor.feature_importances_
feature_names = housing_data.feature_names
```

Normalize the values of the relative feature importance:

```
# Normalize the importance values
feature_importances = 100.0 * (feature_importances /
max(feature_importances))
```

Sort them so that they can be plotted:

```
# Sort the values and flip them
index_sorted = np.flipud(np.argsort(feature_importances))
```

Arrange the ticks on the X axis for the bar graph:

```
# Arrange the X ticks
pos = np.arange(index_sorted.shape[0]) + 0.5
```

Plot the bar graph:

```
# Plot the bar graph
plt.figure()
plt.bar(pos, feature_importances[index_sorted], align='center')
plt.xticks(pos, feature_names[index_sorted])
plt.ylabel('Relative Importance')
plt.title('Feature importance using AdaBoost regressor')
plt.show()
```

The full code is given in the file `feature_importance.py`. If you run the code, you should see the following output:

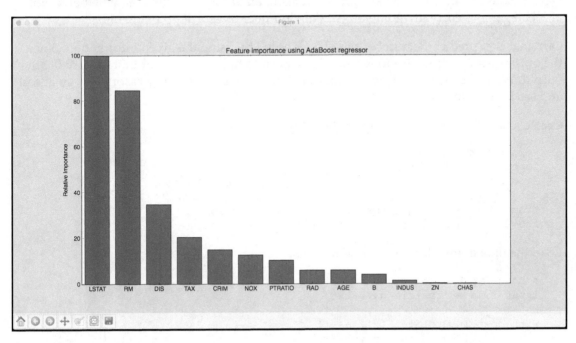

According to this analysis, the feature LSTAT is the most important feature in that dataset.

Predicting traffic using Extremely Random Forest regressor

Let's apply the concepts we learned in the previous sections to a real world problem. We will be using the dataset available at:
`https://archive.ics.uci.edu/ml/datasets/Dodgers+Loop+Sensor`. This dataset consists of data that counts the number of vehicles passing by on the road during baseball games played at Los Angeles Dodgers stadium. In order to make the data readily available for analysis, we need to pre-process it. The pre-processed data is in the file `traffic_data.txt`. In this file, each line contains comma-separated strings. Let's take the first line as an example:

Tuesday,00:00,San Francisco,no,3

With reference to the preceding line, it is formatted as follows:

Day of the week, time of the day, opponent team, binary value indicating whether or not a baseball game is currently going on (yes/no), number of vehicles passing by.

Our goal is to predict the number of vehicles going by using the given information. Since the output variable is continuous valued, we need to build a regressor that can predict the output. We will be using Extremely Random Forests to build this regressor. Let's go ahead and see how to do that.

Create a new Python file and import the following packages:

```
import numpy as np
import matplotlib.pyplot as plt
from sklearn.metrics import classification_report, mean_absolute_error
from sklearn import cross_validation, preprocessing
from sklearn.ensemble import ExtraTreesRegressor
from sklearn.metrics import classification_report
```

Load the data in the file `traffic_data.txt`:

```
# Load input data
input_file = 'traffic_data.txt'
data = []
with open(input_file, 'r') as f:
    for line in f.readlines():
        items = line[:-1].split(',')
        data.append(items)

data = np.array(data)
```

We need to encode the non-numerical features in the data. We also need to ensure that we don't encode numerical features. Each feature that needs to be encoded needs to have a separate label encoder. We need to keep track of these encoders because we will need them when we want to compute the output for an unknown data point. Let's create those label encoders:

```
# Convert string data to numerical data
label_encoder = []
X_encoded = np.empty(data.shape)
for i, item in enumerate(data[0]):
    if item.isdigit():
        X_encoded[:, i] = data[:, i]
    else:
        label_encoder.append(preprocessing.LabelEncoder())
        X_encoded[:, i] = label_encoder[-1].fit_transform(data[:, i])
```

```
X = X_encoded[:, :-1].astype(int)
y = X_encoded[:, -1].astype(int)
```

Split the data into training and testing datasets:

```
# Split data into training and testing datasets
X_train, X_test, y_train, y_test = cross_validation.train_test_split(
        X, y, test_size=0.25, random_state=5)
```

Train an extremely Random Forests regressor:

```
# Extremely Random Forests regressor
params = {'n_estimators': 100, 'max_depth': 4, 'random_state': 0}
regressor = ExtraTreesRegressor(**params)
regressor.fit(X_train, y_train)
```

Compute the performance of the regressor on testing data:

```
# Compute the regressor performance on test data
y_pred = regressor.predict(X_test)
print("Mean absolute error:", round(mean_absolute_error(y_test, y_pred),
2))
```

Let's see how to compute the output for an unknown data point. We will be using those label encoders to convert non-numerical features into numerical values:

```
# Testing encoding on single data instance
test_datapoint = ['Saturday', '10:20', 'Atlanta', 'no']
test_datapoint_encoded = [-1] * len(test_datapoint)
count = 0
for i, item in enumerate(test_datapoint):
        if item.isdigit():
                test_datapoint_encoded[i] = int(test_datapoint[i])
        else:
                test_datapoint_encoded[i] =
int(label_encoder[count].transform(test_datapoint[i]))
                count = count + 1

test_datapoint_encoded = np.array(test_datapoint_encoded)
```

Predict the output:

```
# Predict the output for the test datapoint
print("Predicted traffic:",
    int(regressor.predict([test_datapoint_encoded])[0]))
```

The full code is given in the file `traffic_prediction.py`. If you run the code, you will get 26 as the output, which is pretty close to the actual value. You can confirm this from the data file.

Summary

In this chapter, we learned about Ensemble Learning and how it can be used in the real world. We discussed Decision Trees and how to build a classifier based on it.

We learned about Random Forests and Extremely Random Forests. We discussed how to build classifiers based on them. We understood how to estimate the confidence measure of the predictions. We also learned how to deal with the class imbalance problem.

We discussed how to find the most optimal training parameters to build the models using grid search. We learned how to compute relative feature importance. We then applied ensemble learning techniques to a real-world problem, where we predicted traffic using Extremely Random Forest regressor.

In the next chapter, we will discuss unsupervised learning and how to detect patterns in stock market data.

4
Detecting Patterns with Unsupervised Learning

In this chapter, we are going to learn about unsupervised learning and how to use it in the real world. By the end of this chapter, you will know these things:

- What is unsupervised learning?
- Clustering data with K-Means algorithm
- Estimating the number of clusters with Mean Shift algorithm
- Estimating the quality of clustering with silhouette scores
- What are Gaussian Mixture Models?
- Building a classifier based on Gaussian Mixture Models
- Finding subgroups in stock market using Affinity Propagation model
- Segmenting the market based on shopping patterns

What is unsupervised learning?

Unsupervised learning refers to the process of building machine learning models without using labeled training data. Unsupervised learning finds applications in diverse fields of study, including market segmentation, stock markets, natural language processing, computer vision, and so on.

In the previous chapters, we were dealing with data that had labels associated with it. When we have labeled training data, the algorithms learn to classify data based on those labels. In the real world, we might not always have access to labeled data. Sometimes, we just have a lot of data and we need to categorize it in some way. This is where unsupervised learning comes into picture. Unsupervised learning algorithms attempt to build learning models that can find subgroups within the given dataset using some similarity metric.

Let's see how we formulate the learning problem in unsupervised learning. When we have a dataset without any labels, we assume that the data is generated because of latent variables that govern the distribution in some way. The process of learning can then proceed in a hierarchical manner, starting from the individual data points. We can build deeper levels of representation for the data.

Clustering data with K-Means algorithm

Clustering is one of the most popular unsupervised learning techniques. This technique is used to analyze data and find clusters within that data. In order to find these clusters, we use some kind of similarity measure such as Euclidean distance, to find the subgroups. This similarity measure can estimate the tightness of a cluster. We can say that clustering is the process of organizing our data into subgroups whose elements are similar to each other.

Our goal is to identify the intrinsic properties of data points that make them belong to the same subgroup. There is no universal similarity metric that works for all the cases. It depends on the problem at hand. For example, we might be interested in finding the representative data point for each subgroup or we might be interested in finding the outliers in our data. Depending on the situation, we will end up choosing the appropriate metric.

K-Means algorithm is a well-known algorithm for clustering data. In order to use this algorithm, we need to assume that the number of clusters is known beforehand. We then segment data into K subgroups using various data attributes. We start by fixing the number of clusters and classify our data based on that. The central idea here is that we need to update the locations of these K centroids with each iteration. We continue iterating until we have placed the centroids at their optimal locations.

We can see that the initial placement of centroids plays an important role in the algorithm. These centroids should be placed in a clever manner, because this directly impacts the results. A good strategy is to place them as far away from each other as possible. The basic K-Means algorithm places these centroids randomly where K-Means++ chooses these points algorithmically from the input list of data points. It tries to place the initial centroids far from each other so that it converges quickly. We then go through our training dataset and assign each data point to the closest centroid.

Once we go through the entire dataset, we say that the first iteration is over. We have grouped the points based on the initialized centroids. We now need to recalculate the location of the centroids based on the new clusters that we obtain at the end of the first iteration. Once we obtain the new set of K centroids, we repeat the process again, where we iterate through the dataset and assign each point to the closest centroid.

As we keep repeating these steps, the centroids keep moving to their equilibrium position. After a certain number of iterations, the centroids do not change their locations anymore. This means that we have arrived at the final locations of the centroids. These K centroids are the final K Means that will be used for inference.

Let's apply K-Means clustering on two-dimensional data to see how it works. We will be using the data in the data_clustering.txt file provided to you. Each line contains two comma-separated numbers.

Create a new Python file and import the following packages:

```
import numpy as np
import matplotlib.pyplot as plt
from sklearn.cluster import KMeans
from sklearn import metrics
```

Load the input data from the file:

```
# Load input data
X = np.loadtxt('data_clustering.txt', delimiter=',')
```

We need to define the number of clusters before we can apply K-Means algorithm:

```
num_clusters = 5
```

Visualize the input data to see what the spread looks like:

```
# Plot input data
plt.figure()
plt.scatter(X[:,0], X[:,1], marker='o', facecolors='none',
        edgecolors='black', s=80)
x_min, x_max = X[:, 0].min() - 1, X[:, 0].max() + 1
```

```
y_min, y_max = X[:, 1].min() - 1, X[:, 1].max() + 1
plt.title('Input data')
plt.xlim(x_min, x_max)
plt.ylim(y_min, y_max)
plt.xticks(())
plt.yticks(())
```

We can visually see that there are five groups within this data. Create the `KMeans` object using the initialization parameters. The `init` parameter represents the method of initialization to select the initial centers of clusters. Instead of selecting them randomly, we use k-means++ to select these centers in a smarter way. This ensures that the algorithm converges quickly. The `n_clusters` parameter refers to the number of clusters. The `n_init` parameter refers to the number of times the algorithm should run before deciding upon the best outcome:

```
# Create KMeans object
kmeans = KMeans(init='k-means++', n_clusters=num_clusters, n_init=10)
```

Train the K-Means model with the input data:

```
# Train the KMeans clustering model
kmeans.fit(X)
```

To visualize the boundaries, we need to create a grid of points and evaluate the model on all those points. Let's define the step size of this grid:

```
# Step size of the mesh
step_size = 0.01
```

We define the grid of points and ensure that we are covering all the values in our input data:

```
# Define the grid of points to plot the boundaries
x_min, x_max = X[:, 0].min() - 1, X[:, 0].max() + 1
y_min, y_max = X[:, 1].min() - 1, X[:, 1].max() + 1
x_vals, y_vals = np.meshgrid(np.arange(x_min, x_max, step_size),
         np.arange(y_min, y_max, step_size))
```

Predict the outputs for all the points on the grid using the trained K-Means model:

```
# Predict output labels for all the points on the grid
output = kmeans.predict(np.c_[x_vals.ravel(), y_vals.ravel()])
```

Plot all output values and color each region:

```
# Plot different regions and color them
output = output.reshape(x_vals.shape)
plt.figure()
plt.clf()
plt.imshow(output, interpolation='nearest',
           extent=(x_vals.min(), x_vals.max(),
               y_vals.min(), y_vals.max()),
           cmap=plt.cm.Paired,
           aspect='auto',
           origin='lower')
```

Overlay input data points on top of these colored regions:

```
# Overlay input points
plt.scatter(X[:,0], X[:,1], marker='o', facecolors='none',
        edgecolors='black', s=80)
```

Plot the centers of the clusters obtained using the K-Means algorithm:

```
# Plot the centers of clusters
cluster_centers = kmeans.cluster_centers_
plt.scatter(cluster_centers[:,0], cluster_centers[:,1],
        marker='o', s=210, linewidths=4, color='black',
        zorder=12, facecolors='black')

x_min, x_max = X[:, 0].min() - 1, X[:, 0].max() + 1
y_min, y_max = X[:, 1].min() - 1, X[:, 1].max() + 1
plt.title('Boundaries of clusters')
plt.xlim(x_min, x_max)
plt.ylim(y_min, y_max)
plt.xticks(())
plt.yticks(())
plt.show()
```

The full code is given in the `kmeans.py` file. If you run the code, you will see two screenshot. The first screenshot is the input data:

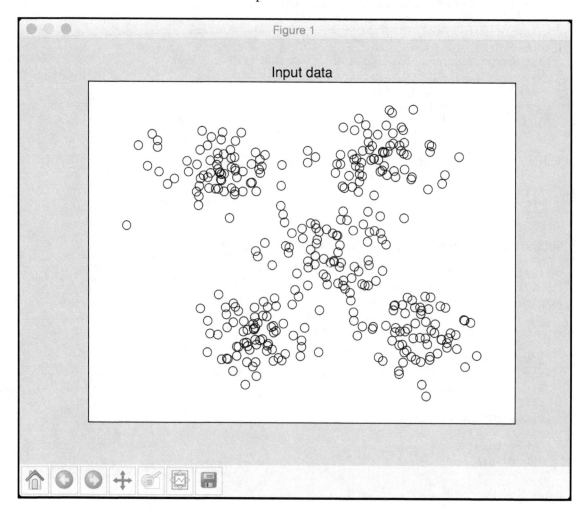

The second screenshot represents the boundaries obtained using K-Means:

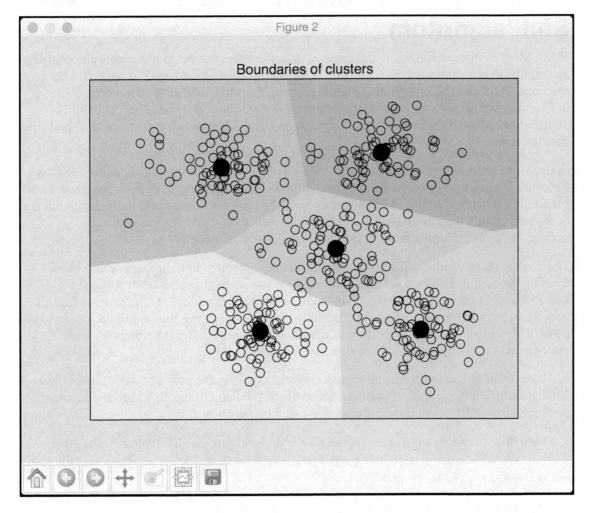

The black filled circle at the center of each cluster represents the centroid of that cluster.

Estimating the number of clusters with Mean Shift algorithm

Mean Shift is a powerful algorithm used in unsupervised learning. It is a non-parametric algorithm used frequently for clustering. It is non-parametric because it does not make any assumptions about the underlying distributions. This is in contrast to parametric techniques, where we assume that the underlying data follows a standard probability distribution. Mean Shift finds a lot of applications in fields like object tracking and real-time data analysis.

In the Mean Shift algorithm, we consider the whole feature space as a probability density function. We start with the training dataset and assume that they have been sampled from a probability density function. In this framework, the clusters correspond to the local maxima of the underlying distribution. If there are K clusters, then there are K peaks in the underlying data distribution and Mean Shift will identify those peaks.

The goal of Mean Shift is to identify the location of centroids. For each data point in the training dataset, it defines a window around it. It then computes the centroid for this window and updates the location to this new centroid. It then repeats the process for this new location by defining a window around it. As we keep doing this, we move closer to the peak of the cluster. Each data point will move towards the cluster it belongs to. The movement is towards a region of higher density.

We keep shifting the centroids, also called means, towards the peaks of each cluster. Since we keep shifting the means, it is called Mean Shift! We keep doing this until the algorithm converges, at which stage the centroids don't move anymore.

Let's see how to use `MeanShift` to estimate the optimal number of clusters in the given dataset. We will be using data in the `data_clustering.txt` file for analysis. It is the same file we used in the *KMeans* section.

Create a new Python file and import the following packages:

```
import numpy as np
import matplotlib.pyplot as plt
from sklearn.cluster import MeanShift, estimate_bandwidth
from itertools import cycle
```

Load input data:

```
# Load data from input file
X = np.loadtxt('data_clustering.txt', delimiter=',')
```

Estimate the bandwidth of the input data. Bandwidth is a parameter of the underlying kernel density estimation process used in Mean Shift algorithm. The bandwidth affects the overall convergence rate of the algorithm and the number of clusters that we will end up with in the end. Hence this is a crucial parameter. If the bandwidth is small, it might results in too many clusters, where as if the value is large, then it will merge distinct clusters.

The `quantile` parameter impacts how the bandwidth is estimated. A higher value for quantile will increase the estimated bandwidth, resulting in a lesser number of clusters:

```
# Estimate the bandwidth of X
bandwidth_X = estimate_bandwidth(X, quantile=0.1, n_samples=len(X))
```

Let's train the Mean Shift clustering model using the estimated bandwidth:

```
# Cluster data with MeanShift
meanshift_model = MeanShift(bandwidth=bandwidth_X, bin_seeding=True)
meanshift_model.fit(X)
```

Extract the centers of all the clusters:

```
# Extract the centers of clusters
cluster_centers = meanshift_model.cluster_centers_
print('\nCenters of clusters:\n', cluster_centers)
```

Extract the number of clusters:

```
# Estimate the number of clusters
labels = meanshift_model.labels_
num_clusters = len(np.unique(labels))
print("\nNumber of clusters in input data =", num_clusters)
```

Visualize the data points:

```
# Plot the points and cluster centers
plt.figure()
markers = 'o*xvs'
for i, marker in zip(range(num_clusters), markers):
    # Plot points that belong to the current cluster
    plt.scatter(X[labels==i, 0], X[labels==i, 1], marker=marker, color='black')
```

Plot the center of the current cluster:

```
    # Plot the cluster center
    cluster_center = cluster_centers[i]
    plt.plot(cluster_center[0], cluster_center[1], marker='o',
            markerfacecolor='black', markeredgecolor='black',
            markersize=15)

plt.title('Clusters')
plt.show()
```

The full code is given in the `mean_shift.py` file. If you run the code, you will see the following screenshot representing the clusters and their centers:

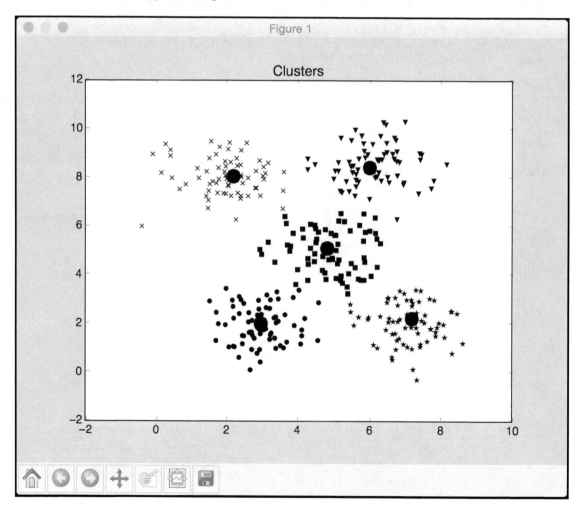

You will see the following on your Terminal:

```
Centers of clusters:
[[ 2.95568966  1.95775862]
 [ 7.17563636  2.18145455]
 [ 2.17603774  8.03283019]
 [ 5.97960784  8.39078431]
 [ 4.81044444  5.07111111]]

Number of clusters in input data = 5
```

Estimating the quality of clustering with silhouette scores

If the data is naturally organized into a number of distinct clusters, then it is easy to visually examine it and draw some inferences. But this is rarely the case in the real world. The data in the real world is huge and messy. So we need a way to quantify the quality of the clustering.

Silhouette refers to a method used to check the consistency of clusters in our data. It gives an estimate of how well each data point fits with its cluster. The silhouette score is a metric that measures how similar a data point is to its own cluster, as compared to other clusters. The silhouette score works with any similarity metric.

For each data point, the silhouette score is computed using the following formula:

silhouette score = (p − q) / max(p, q)

Here, p is the mean distance to the points in the nearest cluster that the data point is not a part of, and q is the mean intra-cluster distance to all the points in its own cluster.

The value of the silhouette score range lies between *-1* to *1*. A score closer to *1* indicates that the data point is very similar to other data points in the cluster, whereas a score closer to *-1* indicates that the data point is not similar to the data points in its cluster. One way to think about it is if you get too many points with negative silhouette scores, then we may have too few or too many clusters in our data. We need to run the clustering algorithm again to find the optimal number of clusters.

Let's see how to estimate the clustering performance using silhouette scores. Create a new Python file and import the following packages:

```
import numpy as np
import matplotlib.pyplot as plt
from sklearn import metrics
from sklearn.cluster import KMeans
```

We will be using the data in the `data_quality.txt` file provided to you. Each line contains two comma-separated numbers:

```
# Load data from input file
X = np.loadtxt('data_quality.txt', delimiter=',')
```

Initialize the variables. The `values` array will contain a list of values we want to iterate on and find the optimal number of clusters:

```
# Initialize variables
scores = []
values = np.arange(2, 10)
```

Iterate through all the values and build a K-Means model during each iteration:

```
# Iterate through the defined range
for num_clusters in values:
    # Train the KMeans clustering model
    kmeans = KMeans(init='k-means++', n_clusters=num_clusters, n_init=10)
    kmeans.fit(X)
```

Estimate the silhouette score for the current clustering model using Euclidean distance metric:

```
score = metrics.silhouette_score(X, kmeans.labels_,
            metric='euclidean', sample_size=len(X))
```

Print the silhouette score for the current value:

```
print("\nNumber of clusters =", num_clusters)
print("Silhouette score =", score)
scores.append(score)
```

Visualize the silhouette scores for various values:

```
# Plot silhouette scores
plt.figure()
plt.bar(values, scores, width=0.7, color='black', align='center')
plt.title('Silhouette score vs number of clusters')
```

Extract the best score and the corresponding value for the number of clusters:

```
# Extract best score and optimal number of clusters
num_clusters = np.argmax(scores) + values[0]
print('\nOptimal number of clusters =', num_clusters)
```

Visualize input data:

```
# Plot data
plt.figure()
plt.scatter(X[:,0], X[:,1], color='black', s=80, marker='o',
        facecolors='none')
x_min, x_max = X[:, 0].min() - 1, X[:, 0].max() + 1
y_min, y_max = X[:, 1].min() - 1, X[:, 1].max() + 1
plt.title('Input data')
plt.xlim(x_min, x_max)
plt.ylim(y_min, y_max)
plt.xticks(())
plt.yticks(())

plt.show()
```

The full code is given in the file `clustering_quality.py`. If you run the code, you will see two screenshot. The first screenshot is the input data:

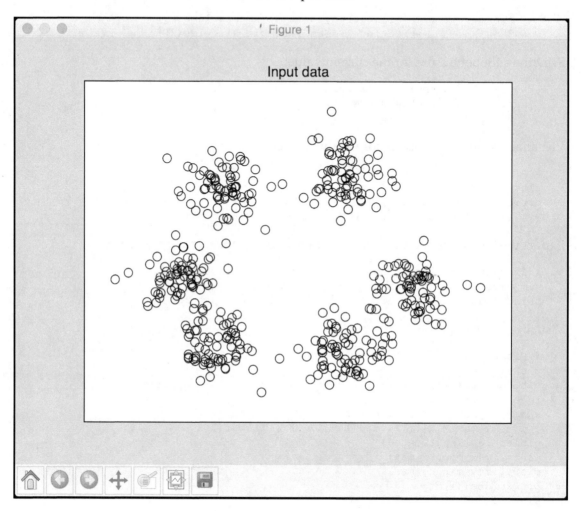

We can see that there are six clusters in our data. The second screenshot represents the scores for various values of number of clusters:

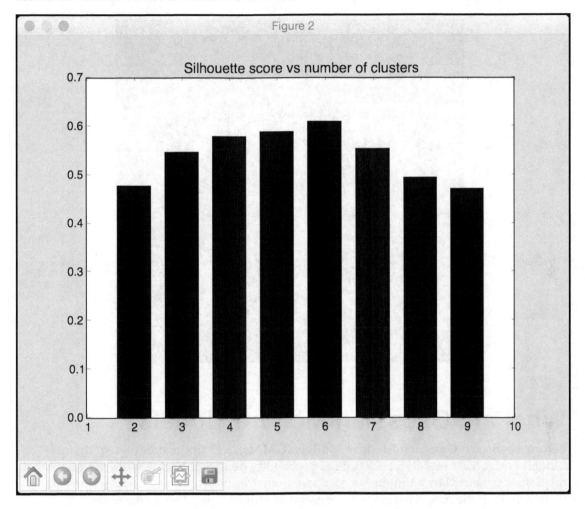

We can verify that the silhouette score is peaking at the value of 6, which is consistent with our data. You will see the following on your Terminal:

```
Number of clusters = 2
Silhouette score = 0.477626248705

Number of clusters = 3
Silhouette score = 0.547174241173

Number of clusters = 4
Silhouette score = 0.579480188969

Number of clusters = 5
Silhouette score = 0.589003263565

Number of clusters = 6
Silhouette score = 0.609690411895

Number of clusters = 7
Silhouette score = 0.554310234032

Number of clusters = 8
Silhouette score = 0.494433661954

Number of clusters = 9
Silhouette score = 0.471414689437

Optimal number of clusters = 6
```

What are Gaussian Mixture Models?

Before we discuss **Gaussian Mixture Models** (**GMM**s), let's understand what Mixture Models are. A Mixture Model is a type of probability density model where we assume that the data is governed by a number of component distributions. If these distributions are Gaussian, then the model becomes a Gaussian Mixture Model. These component distributions are combined in order to provide a multi-modal density function, which becomes a mixture model.

Let's look at an example to understand how Mixture Models work. We want to model the shopping habits of all the people in South America. One way to do it would be model the whole continent and fit everything into a single model. But we know that people in different countries shop differently. We need to understand how people in individual countries shop and how they behave.

If we want to get a good representative model, we need to account for all the variations within the continent. In this case, we can use mixture models to model the shopping habits of individual countries and then combine all of them into a Mixture Model. This way, we are not missing the nuances of the underlying behavior of individual countries. By not enforcing a single model on all the countries, we are able to extract a more accurate model.

An interesting thing to note is that mixture models are semi-parametric, which means that they are partially dependent on a set of predefined functions. They are able to provide greater precision and flexibility in modeling the underlying distributions of our data. They can smooth the gaps that result from having sparse data.

If we define the function, then the mixture model goes from being semi-parametric to parametric. Hence a GMM is a parametric model represented as a weighted summation of component Gaussian functions. We assume that the data is being generated by a set of Gaussian models that are combined in some way. GMMs are very powerful and are used across many fields. The parameters of the GMM are estimated from training data using algorithms like **Expectation–Maximization (EM)** or **Maximum A-Posteriori (MAP)** estimation. Some of the popular applications of GMM include image database retrieval, modeling stock market fluctuations, biometric verification, and so on.

Building a classifier based on Gaussian Mixture Models

Let's build a classifier based on a Gaussian Mixture Model. Create a new Python file and import the following packages:

```
import numpy as np
import matplotlib.pyplot as plt
from matplotlib import patches

from sklearn import datasets
from sklearn.mixture import GMM
from sklearn.cross_validation import StratifiedKFold
```

Let's use the iris dataset available in scikit-learn for analysis:

```
# Load the iris dataset
iris = datasets.load_iris()
```

Split the dataset into training and testing using an 80/20 split. The `n_folds` parameter specifies the number of subsets you'll obtain. We are using a value of 5, which means the dataset will be split into five parts. We will use four parts for training and one part for testing, which gives a split of 80/20:

```
# Split dataset into training and testing (80/20 split)
indices = StratifiedKFold(iris.target, n_folds=5)
```

Extract the training data:

```
# Take the first fold
train_index, test_index = next(iter(indices))

# Extract training data and labels
X_train = iris.data[train_index]
y_train = iris.target[train_index]

# Extract testing data and labels
X_test = iris.data[test_index]
y_test = iris.target[test_index]
```

Extract the number of classes in the training data:

```
# Extract the number of classes
num_classes = len(np.unique(y_train))
```

Build a GMM-based classifier using the relevant parameters. The `n_components` parameter specifies the number of components in the underlying distribution. In this case, it will be the number of distinct classes in our data. We need to specify the type of covariance to use. In this case, we will be using full covariance. The `init_params` parameter controls the parameters that need to be updated during the training process. We have used `wc`, which means weights and covariance parameters will be updated during training. The `n_iter` parameter refers to the number of Expectation-Maximization iterations that will be performed during training:

```
# Build GMM
classifier = GMM(n_components=num_classes, covariance_type='full',
        init_params='wc', n_iter=20)
```

Chapter 4

Initialize the means of the classifier:

```
# Initialize the GMM means
classifier.means_ = np.array([X_train[y_train == i].mean(axis=0)
                    for i in range(num_classes)])
```

Train the Gaussian mixture model classifier using the training data:

```
# Train the GMM classifier
classifier.fit(X_train)
```

Visualize the boundaries of the classifier. We will extract the eigenvalues and eigenvectors to estimate how to draw the elliptical boundaries around the clusters. If you need a quick refresher on eigenvalues and eigenvectors, please refer to: https://www.math.hmc.edu/calculus/tutorials/eigenstuff. Let's go ahead and plot:

```
# Draw boundaries
plt.figure()
colors = 'bgr'
for i, color in enumerate(colors):
    # Extract eigenvalues and eigenvectors
    eigenvalues, eigenvectors = np.linalg.eigh(
            classifier._get_covars()[i][:2, :2])
```

Normalize the first eigenvector:

```
# Normalize the first eigenvector
norm_vec = eigenvectors[0] / np.linalg.norm(eigenvectors[0])
```

The ellipses need to be rotated to accurately show the distribution. Estimate the angle:

```
# Extract the angle of tilt
angle = np.arctan2(norm_vec[1], norm_vec[0])
angle = 180 * angle / np.pi
```

Magnify the ellipses for visualization. The eigenvalues control the size of the ellipses:

```
# Scaling factor to magnify the ellipses
# (random value chosen to suit our needs)
scaling_factor = 8
eigenvalues *= scaling_factor
```

Draw the ellipses:

```
# Draw the ellipse
ellipse = patches.Ellipse(classifier.means_[i, :2],
        eigenvalues[0], eigenvalues[1], 180 + angle,
        color=color)
axis_handle = plt.subplot(1, 1, 1)
ellipse.set_clip_box(axis_handle.bbox)
ellipse.set_alpha(0.6)
axis_handle.add_artist(ellipse)
```

Overlay input data on the figure:

```
# Plot the data
colors = 'bgr'
for i, color in enumerate(colors):
    cur_data = iris.data[iris.target == i]
    plt.scatter(cur_data[:,0], cur_data[:,1], marker='o',
            facecolors='none', edgecolors='black', s=40,
            label=iris.target_names[i])
```

Overlay test data on this figure:

```
    test_data = X_test[y_test == i]
    plt.scatter(test_data[:,0], test_data[:,1], marker='s',
            facecolors='black', edgecolors='black', s=40,
            label=iris.target_names[i])
```

Compute the predicted output for training and testing data:

```
# Compute predictions for training and testing data
y_train_pred = classifier.predict(X_train)
accuracy_training = np.mean(y_train_pred.ravel() == y_train.ravel()) * 100
print('Accuracy on training data =', accuracy_training)

y_test_pred = classifier.predict(X_test)
accuracy_testing = np.mean(y_test_pred.ravel() == y_test.ravel()) * 100
print('Accuracy on testing data =', accuracy_testing)

plt.title('GMM classifier')
plt.xticks(())
plt.yticks(())

plt.show()
```

The full code is given in the file `gmm_classifier.py`. If you run the code, you will see the following output:

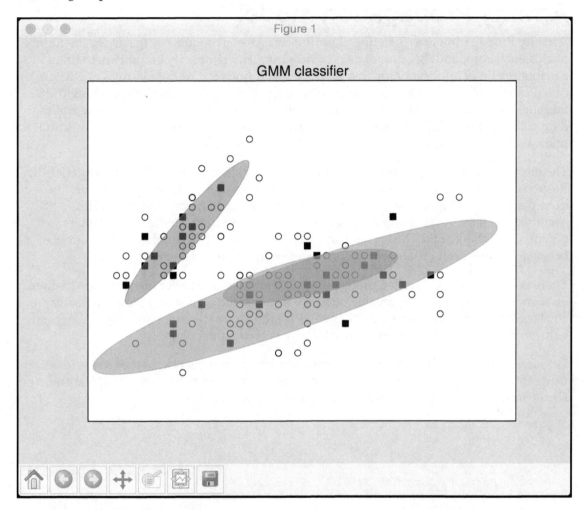

The input data consists of three distributions. The three ellipses of various sizes and angles represent the underlying distributions in the input data. You will see the following printed on your Terminal:

```
Accuracy on training data = 87.5
Accuracy on testing data = 86.6666666667
```

Finding subgroups in stock market using Affinity Propagation model

Affinity Propagation is a clustering algorithm that doesn't require us to specify the number of clusters beforehand. Because of its generic nature and simplicity of implementation, it has found a lot of applications across many fields. It finds out representatives of clusters, called exemplars, using a technique called message passing. We start by specifying the measures of similarity that we want it to consider. It simultaneously considers all training data points as potential exemplars. It then passes messages between the data points until it finds a set of exemplars.

The message passing happens in two alternate steps, called **responsibility** and **availability.** Responsibility refers to the message sent from members of the cluster to candidate exemplars, indicating how well suited the data point would be as a member of this exemplar's cluster. Availability refers to the message sent from candidate exemplars to potential members of the cluster, indicating how well suited it would be as an exemplar. It keeps doing this until the algorithm converges on an optimal set of exemplars.

There is also a parameter called preference that controls the number of exemplars that will be found. If you choose a high value, then it will cause the algorithm to find too many clusters. If you choose a low value, then it will lead to a small number of clusters. A good value to choose would be the median similarity between the points.

Let's use Affinity Propagation model to find subgroups in the stock market. We will be using the stock quote variation between opening and closing as the governing feature. Create a new Python file and import the following packages:

```
import datetime
import json

import numpy as np
import matplotlib.pyplot as plt
from sklearn import covariance, cluster
from matplotlib.finance import quotes_historical_yahoo_ochl as quotes_yahoo
```

We will be using the stock market data available in `matplotlib`. The company symbols are mapped to their full names in the file `company_symbol_mapping.json`:

```
# Input file containing company symbols
input_file = 'company_symbol_mapping.json'
```

Load the company symbol map from the file:

```
# Load the company symbol map
with open(input_file, 'r') as f:
    company_symbols_map = json.loads(f.read())

symbols, names = np.array(list(company_symbols_map.items())).T
```

Load the stock quotes from matplotlib:

```
# Load the historical stock quotes
start_date = datetime.datetime(2003, 7, 3)
end_date = datetime.datetime(2007, 5, 4)
quotes = [quotes_yahoo(symbol, start_date, end_date, asobject=True)
                for symbol in symbols]
```

Compute the difference between opening and closing quotes:

```
# Extract opening and closing quotes
opening_quotes = np.array([quote.open for quote in
quotes]).astype(np.float)
closing_quotes = np.array([quote.close for quote in
quotes]).astype(np.float)

# Compute differences between opening and closing quotes
quotes_diff = closing_quotes - opening_quotes
```

Normalize the data:

```
# Normalize the data
X = quotes_diff.copy().T
X /= X.std(axis=0)
```

Create a graph model:

```
# Create a graph model
edge_model = covariance.GraphLassoCV()
```

Train the model:

```
# Train the model
with np.errstate(invalid='ignore'):
    edge_model.fit(X)
```

Detecting Patterns with Unsupervised Learning

Build the affinity propagation clustering model using the edge model we just trained:

```
# Build clustering model using Affinity Propagation model
_, labels = cluster.affinity_propagation(edge_model.covariance_)
num_labels = labels.max()
```

Print the output:

```
# Print the results of clustering
for i in range(num_labels + 1):
    print("Cluster", i+1, "==>", ', '.join(names[labels == i]))
```

The full code is given in the file `stocks.py`. If you run the code, you will see the following output on your Terminal:

```
Clustering of stocks based on difference in opening and closing quotes:

Cluster 1 ==> Kraft Foods
Cluster 2 ==> CVS, Walgreen
Cluster 3 ==> Amazon, Yahoo
Cluster 4 ==> Cablevision
Cluster 5 ==> Pfizer, Sanofi-Aventis, GlaxoSmithKline, Novartis
Cluster 6 ==> HP, General Electrics, 3M, Microsoft, Cisco, IBM, Texas instruments, Dell
Cluster 7 ==> Coca Cola, Kimberly-Clark, Pepsi, Procter Gamble, Kellogg, Colgate-Palmolive
Cluster 8 ==> Comcast, Wells Fargo, Xerox, Home Depot, Wal-Mart, Marriott, Navistar, DuPont de Nemours, American express, Ryder, JPMorgan Chase, AIG, Time Warner, Bank of America, Goldman Sachs
Cluster 9 ==> Canon, Unilever, Mitsubishi, Apple, Mc Donalds, Boeing, Toyota, Caterpillar, Ford, Honda, SAP, Sony
Cluster 10 ==> Valero Energy, Exxon, ConocoPhillips, Chevron, Total
Cluster 11 ==> Raytheon, General Dynamics, Lookheed Martin, Northrop Grumman
```

This output represents the various subgroups in the stock market during that time period. Please note that the clusters might appear in a different order when you run the code.

Segmenting the market based on shopping patterns

Let's see how to apply unsupervised learning techniques to segment the market based on customer shopping habits. You have been provided with a file named `sales.csv`. This file contains the sales details of a variety of tops from a number of retail clothing stores. Our goal is to identify the patterns and segment the market based on the number of units sold in these stores.

Chapter 4

Create a new Python file and import the following packages:

```
import csv

import numpy as np
import matplotlib.pyplot as plt
from sklearn.cluster import MeanShift, estimate_bandwidth
```

Load the data from the input file. Since it's a csv file, we can use the csv reader in python to read the data from this file and convert it into a NumPy array:

```
# Load data from input file
input_file = 'sales.csv'
file_reader = csv.reader(open(input_file, 'r'), delimiter=',')

X = []
for count, row in enumerate(file_reader):
    if not count:
        names = row[1:]
        continue

    X.append([float(x) for x in row[1:]])

# Convert to numpy array
X = np.array(X)
```

Let's estimate the bandwidth of the input data:

```
# Estimating the bandwidth of input data
bandwidth = estimate_bandwidth(X, quantile=0.8, n_samples=len(X))
```

Train a mean shift model based on the estimated bandwidth:

```
# Compute clustering with MeanShift
meanshift_model = MeanShift(bandwidth=bandwidth, bin_seeding=True)
meanshift_model.fit(X)
```

Extract the labels and the centers of each cluster:

```
labels = meanshift_model.labels_
cluster_centers = meanshift_model.cluster_centers_
num_clusters = len(np.unique(labels))
```

Print the number of clusters and the cluster centers:

```
print("\nNumber of clusters in input data =", num_clusters)

print("\nCenters of clusters:")
print('\t'.join([name[:3] for name in names]))
for cluster_center in cluster_centers:
    print('\t'.join([str(int(x)) for x in cluster_center]))
```

We are dealing with six-dimensional data. In order to visualize the data, let's take two-dimensional data formed using second and third dimensions:

```
# Extract two features for visualization
cluster_centers_2d = cluster_centers[:, 1:3]
```

Plot the centers of clusters:

```
# Plot the cluster centers
plt.figure()
plt.scatter(cluster_centers_2d[:,0], cluster_centers_2d[:,1],
        s=120, edgecolors='black', facecolors='none')

offset = 0.25
plt.xlim(cluster_centers_2d[:,0].min() - offset *
cluster_centers_2d[:,0].ptp(),
        cluster_centers_2d[:,0].max() + offset *
cluster_centers_2d[:,0].ptp(),)
plt.ylim(cluster_centers_2d[:,1].min() - offset *
cluster_centers_2d[:,1].ptp(),
        cluster_centers_2d[:,1].max() + offset *
cluster_centers_2d[:,1].ptp())

plt.title('Centers of 2D clusters')
plt.show()
```

Chapter 4

The full code is given in the file market_segmentation.py. If you run the code, you will see the following output:

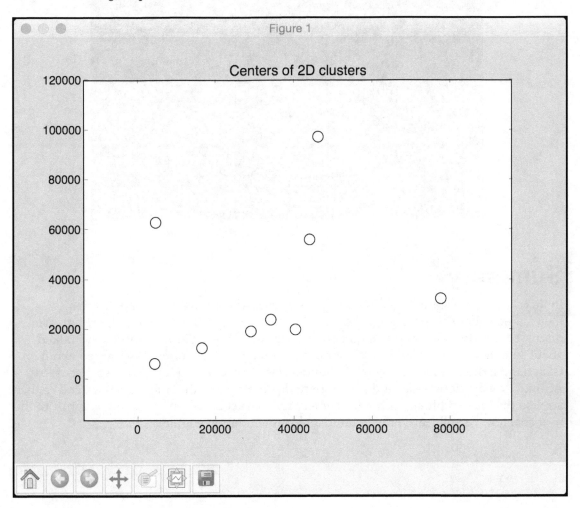

You will see the following on your Terminal:

```
Number of clusters in input data = 9

Centers of clusters:
Tsh      Tan      Hal      Tur      Tub      Swe
9823     4637     6539     2607     2228     1239
38589    44199    56158    5030     24674    4125
7852     4939     63081    134      40066    1332
35314    16745    12775    66900    1298     5613
22617    77873    32543    1005     21035    837
104972   29186    19415    16016    5060     9372
38741    40539    20120    35059    255      50710
28333    34263    24065    5575     4229     18076
14987    46397    97393    1127     37315    3235
```

Summary

In this chapter, we started by discussing unsupervised learning and its applications. We then learned about clustering and how to cluster data using the K-Means algorithm. We discussed how to estimate the number of clusters with Mean Shift algorithm. We talked about silhouette scores and how to estimate the quality of clustering. We learned about Gaussian Mixture Models and how to build a classifier based on that. We also discussed Affinity Propagation model and used it to find subgroups within the stock market. We then applied the Mean Shift algorithm to segment the market based on shopping patterns. In the next chapter, we will learn how to build a recommendation engine.

5
Building Recommender Systems

In this chapter, we are going to learn how to build a movie recommendation system. We will discuss how to create a training pipeline that can be trained with custom parameters. We will then learn about the Nearest Neighbors classifier and see how to implement it. We use these concepts to discuss collaborative filtering and then use it to build a recommender system.

By the end of this chapter, you will learn about the following:

- Creating a training pipeline
- Extracting the nearest neighbors
- Building a K Nearest Neighbors classifier
- Computing similarity scores
- Finding similar users using collaborative filtering
- Building a movie recommendation system

Creating a training pipeline

Machine-learning systems are usually built using different modules. These modules are combined in a particular way to achieve an end goal. The `scikit-learn` library has functions that enable us to build these pipelines by concatenating various modules together. We just need to specify the modules along with the corresponding parameters. It will then build a pipeline using these modules that processes the data and trains the system.

The pipeline can include modules that perform various functions like feature selection, preprocessing, random forests, clustering, and so on. In this section, we will see how to build a pipeline to select the top K features from an input data point and then classify them using an Extremely Random Forest classifier.

Create a new Python file and import the following packages:

```
from sklearn.datasets import samples_generator
from sklearn.feature_selection import SelectKBest, f_regression
from sklearn.pipeline import Pipeline
from sklearn.ensemble import ExtraTreesClassifier
```

Let's generate some labeled sample data for training and testing. The `scikit-learn` package has a built-in function that handles it. In the line to follow, we create 150 data points, where each data point is a 25-dimensional feature vector. The numbers in each feature vector will be generated using a random sample generator. Each data point has six informative features and no redundant features. Use the following code:

```
# Generate data
X, y = samples_generator.make_classification(n_samples=150,
        n_features=25, n_classes=3, n_informative=6,
        n_redundant=0, random_state=7)
```

The first block in the pipeline is the feature selector. This block selects the K best features. Let's set the value of K to 9, as follows:

```
# Select top K features
k_best_selector = SelectKBest(f_regression, k=9)
```

The next block in the pipeline is an Extremely Random Forests classifier with 60 estimators and a maximum depth of four. Use the following code:

```
# Initialize Extremely Random Forests classifier
classifier = ExtraTreesClassifier(n_estimators=60, max_depth=4)
```

Let's construct the pipeline by joining the individual blocks that we've constructed. We can name each block so that it's easier to track:

```
# Construct the pipeline
processor_pipeline = Pipeline([('selector', k_best_selector), ('erf',
classifier)])
```

We can change the parameters of the individual blocks. Let's change the value of K in the first block to 7 and the number of estimators in the second block to 30. We will use the names we assigned in the previous line to define the scope:

```
# Set the parameters
processor_pipeline.set_params(selector__k=7, erf__n_estimators=30)
```

Train the pipeline using the sample data that we generated earlier:

```
# Training the pipeline
processor_pipeline.fit(X, y)
```

Predict the output for all the input values and print it:

```
# Predict outputs for the input data
output = processor_pipeline.predict(X)
print("\nPredicted output:\n", output)
```

Compute the score using the labeled training data:

```
# Print scores
print("\nScore:", processor_pipeline.score(X, y))
```

Extract the features chosen by the selector block. We specified that we wanted to choose 7 features out of 25. Use the following code:

```
# Print the features chosen by the pipeline selector
status = processor_pipeline.named_steps['selector'].get_support()

# Extract and print indices of selected features
selected = [i for i, x in enumerate(status) if x]
print("\nIndices of selected features:", ', '.join([str(x) for x in selected]))
```

The full code is given in the file `pipeline_trainer.py`. If you run the code, you will see the following output on your Terminal:

```
Predicted output:
[1 2 2 0 2 2 0 2 0 1 2 0 2 1 0 0 2 2 2 1 0 2 0 1 2 1 1 1 0 0 1 2 1 0 0 0 2
 1 1 0 2 0 0 0 1 2 0 2 1 0 1 0 0 0 2 1 1 1 1 1 0 1 2 2 2 0 2 0 2 2 0 1 2 0
 2 0 2 0 1 0 2 2 1 1 1 2 0 0 0 0 2 2 0 2 1 1 2 0 1 1 2 1 1 0 1 0 2 2 2 0 0
 1 2 1 1 0 2 0 0 0 0 2 2 1 1 1 2 0 2 2 1 0 2 0 0 0 1 1 2 2 2 2 2 1 1 0
 2 0]

Score: 0.893333333333

Indices of selected features: 13, 15, 18, 19, 21, 23, 24
```

The predicted output list in the preceding screenshot shows the output labels predicted using the processor. The score represents the effectiveness of the processor. The last line indicates the indices of the chosen features.

Extracting the nearest neighbors

Recommender systems employ the concept of nearest neighbors to find good recommendations. Nearest neighbors refers to the process of finding the closest points to the input point from the given dataset. This is frequently used to build classification systems that classify a datapoint based on the proximity of the input data point to various classes. Let's see how to find the nearest neighbors of a given data point.

Create a new Python file and import the following packages:

```
import numpy as np
import matplotlib.pyplot as plt
from sklearn.neighbors import NearestNeighbors
```

Define sample 2D datapoints:

```
# Input data
X = np.array([[2.1, 1.3], [1.3, 3.2], [2.9, 2.5], [2.7, 5.4], [3.8, 0.9],
              [7.3, 2.1], [4.2, 6.5], [3.8, 3.7], [2.5, 4.1], [3.4, 1.9],
              [5.7, 3.5], [6.1, 4.3], [5.1, 2.2], [6.2, 1.1]])
```

Define the number of nearest neighbors you want to extract:

```
# Number of nearest neighbors
k = 5
```

Define a test datapoint that will be used to extract the K nearest neighbors:

```
# Test datapoint
test_datapoint = [4.3, 2.7]
```

Plot the input data using circular shaped black markers:

```
# Plot input data
plt.figure()
plt.title('Input data')
plt.scatter(X[:,0], X[:,1], marker='o', s=75, color='black')
```

Create and train a **K Nearest Neighbors** model using the input data. Use this model to extract the nearest neighbors to our test data point:

```
# Build K Nearest Neighbors model
knn_model = NearestNeighbors(n_neighbors=k, algorithm='ball_tree').fit(X)
distances, indices = knn_model.kneighbors(test_datapoint)
```

Print the nearest neighbors extracted from the model:

```
# Print the 'k' nearest neighbors
print("\nK Nearest Neighbors:")
for rank, index in enumerate(indices[0][:k], start=1):
    print(str(rank) + " ==>", X[index])
```

Visualize the nearest neighbors:

```
# Visualize the nearest neighbors along with the test datapoint
plt.figure()
plt.title('Nearest neighbors')
plt.scatter(X[:, 0], X[:, 1], marker='o', s=75, color='k')
plt.scatter(X[indices][0][:][:, 0], X[indices][0][:][:, 1],
        marker='o', s=250, color='k', facecolors='none')
plt.scatter(test_datapoint[0], test_datapoint[1],
        marker='x', s=75, color='k')

plt.show()
```

The full code is given in the file `k_nearest_neighbors.py`. If you run the code, you will see two screenshot. The first screenshot represents the input data:

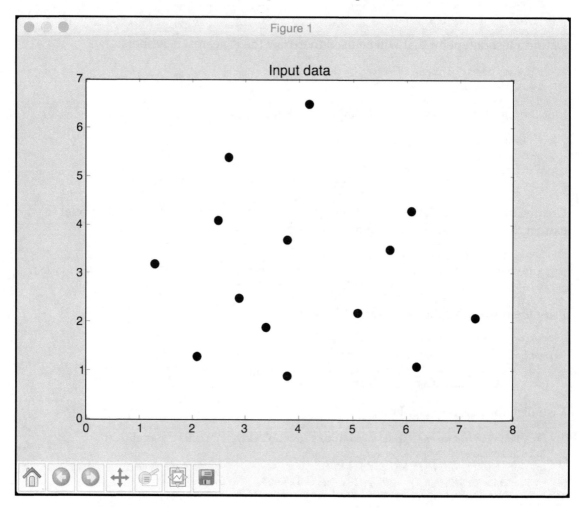

The second screenshot represents the five nearest neighbors. The test data point is shown using a cross and the nearest neighbor points have been circled:

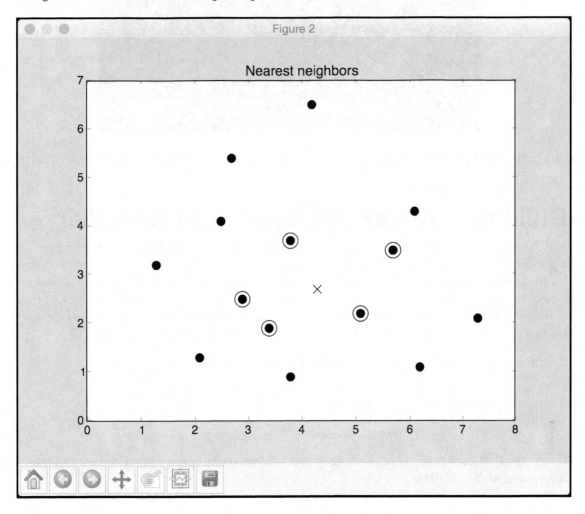

You will see the following output on your Terminal:

```
K Nearest Neighbors:
1 ==> [ 5.1  2.2]
2 ==> [ 3.8  3.7]
3 ==> [ 3.4  1.9]
4 ==> [ 2.9  2.5]
5 ==> [ 5.7  3.5]
```

The preceding figure shows the five points that are closest to the test data point.

Building a K-Nearest Neighbors classifier

A K-Nearest Neighbors classifier is a classification model that uses the nearest neighbors algorithm to classify a given data point. The algorithm finds the K closest data points in the training dataset to identify the category of the input data point. It will then assign a class to this data point based on a majority vote. From the list of those K data points, we look at the corresponding classes and pick the one with the highest number of votes. Let's see how to build a classifier using this model. The value of K depends on the problem at hand.

Create a new Python file and import the following packages:

```
import numpy as np
import matplotlib.pyplot as plt
import matplotlib.cm as cm
from sklearn import neighbors, datasets
```

Load the input data from data.txt. Each line contains comma-separated values and the data contains four classes:

```
# Load input data
input_file = 'data.txt'
data = np.loadtxt(input_file, delimiter=',')
X, y = data[:, :-1], data[:, -1].astype(np.int)
```

Visualize the input data using four different marker shapes. We need to map the labels to corresponding markers, which is where the mapper variable comes into the picture:

```
# Plot input data
plt.figure()
plt.title('Input data')
```

```
marker_shapes = 'v^os'
mapper = [marker_shapes[i] for i in y]
for i in range(X.shape[0]):
    plt.scatter(X[i, 0], X[i, 1], marker=mapper[i],
            s=75, edgecolors='black', facecolors='none')
```

Define the number of nearest neighbors we want to use:

```
# Number of nearest neighbors
num_neighbors = 12
```

Define the step size of the grid that will be used to visualize the boundaries of the classifier model:

```
# Step size of the visualization grid
step_size = 0.01
```

Create the K Nearest Neighbors classifier model:

```
# Create a K Nearest Neighbors classifier model
classifier = neighbors.KNeighborsClassifier(num_neighbors,
weights='distance')
```

Train the model using training data:

```
# Train the K Nearest Neighbours model
classifier.fit(X, y)
```

Create the mesh grid of values that will be used to visualize the grid:

```
# Create the mesh to plot the boundaries
x_min, x_max = X[:, 0].min() - 1, X[:, 0].max() + 1
y_min, y_max = X[:, 1].min() - 1, X[:, 1].max() + 1
x_values, y_values = np.meshgrid(np.arange(x_min, x_max, step_size),
        np.arange(y_min, y_max, step_size))
```

Evaluate the classifier on all the points on the grid to create a visualization of the boundaries:

```
# Evaluate the classifier on all the points on the grid
output = classifier.predict(np.c_[x_values.ravel(), y_values.ravel()])
```

Create a color mesh to visualize the output:

```
# Visualize the predicted output
output = output.reshape(x_values.shape)
plt.figure()
plt.pcolormesh(x_values, y_values, output, cmap=cm.Paired)
```

Building Recommender Systems

Overlay training data on top of this color mesh to visualize the data relative to the boundaries:

```
# Overlay the training points on the map
for i in range(X.shape[0]):
    plt.scatter(X[i, 0], X[i, 1], marker=mapper[i],
            s=50, edgecolors='black', facecolors='none')
```

Set the X and Y limits along with the title:

```
plt.xlim(x_values.min(), x_values.max())
plt.ylim(y_values.min(), y_values.max())
plt.title('K Nearest Neighbors classifier model boundaries')
```

Define a test datapoint to see how the classifier performs. Create a figure with training data points and a test data point to see where it lies:

```
# Test input datapoint
test_datapoint = [5.1, 3.6]
plt.figure()
plt.title('Test datapoint')
for i in range(X.shape[0]):
    plt.scatter(X[i, 0], X[i, 1], marker=mapper[i],
            s=75, edgecolors='black', facecolors='none')

plt.scatter(test_datapoint[0], test_datapoint[1], marker='x',
        linewidth=6, s=200, facecolors='black')
```

Extract the K Nearest Neighbors to the test data point, based on the classifier model:

```
# Extract the K nearest neighbors
_, indices = classifier.kneighbors([test_datapoint])
indices = indices.astype(np.int)[0]
```

Plot the K nearest neighbors obtained in the previous step:

```
# Plot k nearest neighbors
plt.figure()
plt.title('K Nearest Neighbors')

for i in indices:
    plt.scatter(X[i, 0], X[i, 1], marker=mapper[y[i]],
            linewidth=3, s=100, facecolors='black')
```

Overlay the test data point:

```
plt.scatter(test_datapoint[0], test_datapoint[1], marker='x',
        linewidth=6, s=200, facecolors='black')
```

Overlay the input data:

```
for i in range(X.shape[0]):
    plt.scatter(X[i, 0], X[i, 1], marker=mapper[i],
            s=75, edgecolors='black', facecolors='none')
```

Print the predicted output:

```
print("Predicted output:", classifier.predict([test_datapoint])[0])

plt.show()
```

The full code is given in the file nearest_neighbors_classifier.py. If you run the code, you will see four screenshot. The first screenshot represents the input data:

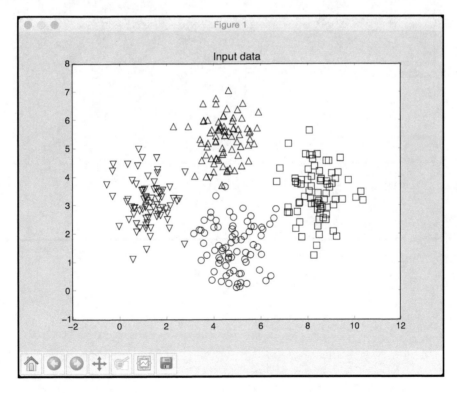

The second screenshot represents the classifier boundaries:

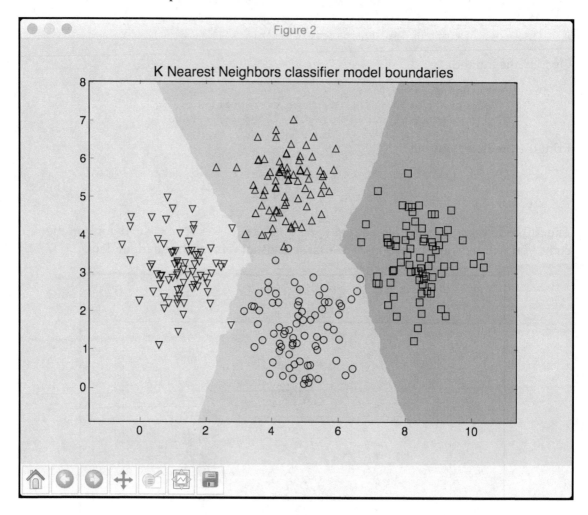

The third screenshot shows the test data point relative to the input dataset. The test data point is shown using a cross:

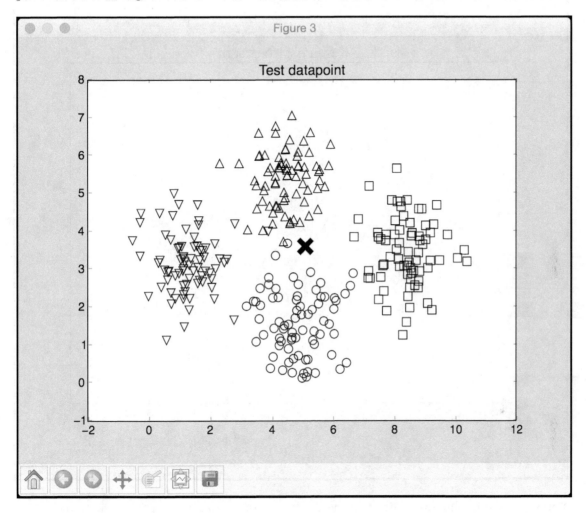

The fourth screenshot shows the 12 nearest neighbors to the test data point:

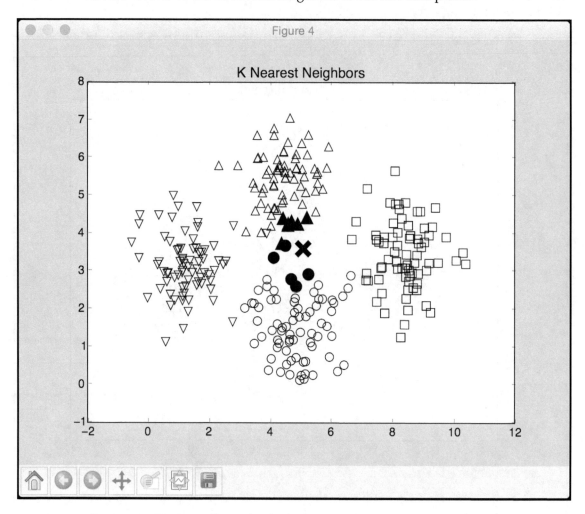

You will see the following output on the Terminal, indicating that the test data point belongs to class 1:

```
Predicted output: 1
```

Computing similarity scores

In order to build a recommendation system, it is important to understand how to compare various objects in our dataset. Let's say our dataset consists of people and their various movie preferences. In order to recommend something, we need to understand how to compare any two people with each other. This is where the similarity score becomes very important. The similarity score gives us an idea of how similar two objects are.

There are two scores that are used frequently in this domain — Euclidean score and Pearson score. **Euclidean score** uses the Euclidean distance between two data points to compute the score. If you need a quick refresher on how Euclidean distance is computed, you can go to https://en.wikipedia.org/wiki/Euclidean_distance. The value of the Euclidean distance can be unbounded. Hence we take this value and convert it in a way that the Euclidean score ranges from 0 to 1. If the Euclidean distance between two objects is large, then the Euclidean score should be low because a low score indicates that the objects are not similar. Hence Euclidean distance is inversely proportional to Euclidean score.

Pearson score is a measure of correlation between two objects. It uses the covariance between the two objects along with their individual standard deviations to compute the score. The score can range from -1 to +1. A score of +1 indicates that the objects are very similar where a score of -1 would indicate that the objects are very dissimilar. A score of 0 would indicate that there is no correlation between the two objects. Let's see how to compute these scores.

Create a new Python file and import the following packages:

```
import argparse
import json
import numpy as np
```

Build an argument parser to process the input arguments. It will accept two users and the type of score that it needs to use to compute the similarity score:

```
def build_arg_parser():
    parser = argparse.ArgumentParser(description='Compute similarity score')
    parser.add_argument('--user1', dest='user1', required=True,
            help='First user')
    parser.add_argument('--user2', dest='user2', required=True,
            help='Second user')
    parser.add_argument("--score-type", dest="score_type", required=True,
            choices=['Euclidean', 'Pearson'], help='Similarity metric to be used')
    return parser
```

Define a function to compute the Euclidean score between the input users. If the users are not in the dataset, raise an error:

```
# Compute the Euclidean distance score between user1 and user2
def euclidean_score(dataset, user1, user2):
    if user1 not in dataset:
        raise TypeError('Cannot find ' + user1 + ' in the dataset')

    if user2 not in dataset:
        raise TypeError('Cannot find ' + user2 + ' in the dataset')
```

Define a variable to track the movies that have been rated by both the users:

```
# Movies rated by both user1 and user2
common_movies = {}
```

Extract the movies rated by both users:

```
for item in dataset[user1]:
    if item in dataset[user2]:
        common_movies[item] = 1
```

If there are no common movies, then we cannot compute the similarity score:

```
# If there are no common movies between the users,
# then the score is 0
if len(common_movies) == 0:
    return 0
```

Compute the squared differences between the ratings and use it to compute the Euclidean score:

```
squared_diff = []

for item in dataset[user1]:
    if item in dataset[user2]:
        squared_diff.append(np.square(dataset[user1][item] - dataset[user2][item]))
    return 1 / (1 + np.sqrt(np.sum(squared_diff)))
```

Define a function to compute the Pearson score between the input users in the given dataset. If the users are not found in the dataset, raise an error:

```
# Compute the Pearson correlation score between user1 and user2
def pearson_score(dataset, user1, user2):
    if user1 not in dataset:
        raise TypeError('Cannot find ' + user1 + ' in the dataset')
```

```
        if user2 not in dataset:
            raise TypeError('Cannot find ' + user2 + ' in the dataset')
```

Define a variable to track the movies that have been rated by both the users:

```
    # Movies rated by both user1 and user2
    common_movies = {}
```

Extract the movies rated by both users:

```
        for item in dataset[user1]:
            if item in dataset[user2]:
                common_movies[item] = 1
```

If there are no common movies, then we cannot compute the similarity score:

```
    num_ratings = len(common_movies)

    # If there are no common movies between user1 and user2, then the score is 0
    if num_ratings == 0:
        return 0
```

Calculate the sum of ratings of all the movies that have been rated by both the users:

```
    # Calculate the sum of ratings of all the common movies
    user1_sum = np.sum([dataset[user1][item] for item in common_movies])
    user2_sum = np.sum([dataset[user2][item] for item in common_movies])
```

Calculate the sum of squares of the ratings all the movies that have been rated by both the users:

```
    # Calculate the sum of squares of ratings of all the common movies
    user1_squared_sum = np.sum([np.square(dataset[user1][item]) for item in common_movies])
    user2_squared_sum = np.sum([np.square(dataset[user2][item]) for item in common_movies])
```

Calculate the sum of products of the ratings of all the movies rated by both the input users:

```
    # Calculate the sum of products of the ratings of the common movies
    sum_of_products = np.sum([dataset[user1][item] * dataset[user2][item] for item in common_movies])
```

Calculate the various parameters required to compute the Pearson score using the preceding computations:

```
# Calculate the Pearson correlation score
Sxy = sum_of_products - (user1_sum * user2_sum / num_ratings)
Sxx = user1_squared_sum - np.square(user1_sum) / num_ratings
Syy = user2_squared_sum - np.square(user2_sum) / num_ratings
```

If there is no deviation, then the score is 0:

```
if Sxx * Syy == 0:
    return 0
```

Return the Pearson score:

```
return Sxy / np.sqrt(Sxx * Syy)
```

Define the main function and parse the input arguments:

```
if __name__=='__main__':
    args = build_arg_parser().parse_args()
    user1 = args.user1
    user2 = args.user2
    score_type = args.score_type
```

Load the ratings from the file `ratings.json` into a dictionary:

```
ratings_file = 'ratings.json'

with open(ratings_file, 'r') as f:
    data = json.loads(f.read())
```

Compute the similarity score based on the input arguments:

```
if score_type == 'Euclidean':
    print("\nEuclidean score:")
    print(euclidean_score(data, user1, user2))
else:
    print("\nPearson score:")
    print(pearson_score(data, user1, user2))
```

The full code is given in the file `compute_scores.py`. Let's run the code with a few combinations. Let's say we want to compute the Euclidean score between `David Smith` and `Bill Duffy`:

```
$ python3 compute_scores.py --user1 "David Smith" --user2 "Bill Duffy"
--score-type Euclidean
```

If you run the above command, you will get the following output on your Terminal:

```
Euclidean score:
0.585786437627
```

If you want to compute the Pearson score between the same pair, run the following command on your Terminal:

```
$ python3 compute_scores.py --user1 "David Smith" --user2 "Bill Duffy" --score-type Pearson
```

You will see the following on your Terminal:

```
Pearson score:
0.99099243041
```

You can run it using other combinations of parameters as well.

Finding similar users using collaborative filtering

Collaborative filtering refers to the process of identifying patterns among the objects in a dataset in order to make a decision about a new object. In the context of recommendation engines, we use collaborative filtering to provide recommendations by looking at similar users in the dataset.

By collecting the preferences of different users in the dataset, we collaborate that information to filter the users. Hence the name collaborative filtering.

The assumption here is that if two people have similar ratings for a particular set of movies, then their choices in a set of new unknown movies would be similar too. By identifying patterns in those common movies, we make predictions about new movies. In the previous section, we learned how to compare different users in the dataset. We will use these scoring techniques to find similar users in our dataset. Collaborative filtering is typically used when we have huge datasets. These methods can be used for various verticals like finance, online shopping, marketing, customer studies, and so on.

Create a new Python file and import the following packages:

```
import argparse
import json
import numpy as np

from compute_scores import pearson_score
```

Define a function to parse the input arguments. The only input argument would be the name of the user:

```
def build_arg_parser():
    parser = argparse.ArgumentParser(description='Find users who are similar to the input user ')
    parser.add_argument('--user', dest='user', required=True,
            help='Input user')
    return parser
```

Define a function to find the users in the dataset that are similar to the given user. If the user is not in the dataset, raise an error:

```
# Finds users in the dataset that are similar to the input user
def find_similar_users(dataset, user, num_users):
    if user not in dataset:
        raise TypeError('Cannot find ' + user + ' in the dataset')
```

We have already imported the function to compute the Pearson score. Let's use that function to compute the Pearson score between the input user and all the other users in the dataset:

```
# Compute Pearson score between input user
# and all the users in the dataset
scores = np.array([[x, pearson_score(dataset, user,
        x)] for x in dataset if x != user])
```

Sort the scores in descending order:

```
# Sort the scores in decreasing order
scores_sorted = np.argsort(scores[:, 1])[::-1]
```

Extract the top `num_users` number of users as specified by the input argument and return the array:

```
# Extract the top 'num_users' scores
top_users = scores_sorted[:num_users]
return scores[top_users]
```

Define the main function and parse the input arguments to extract the name of the user:

```
if __name__=='__main__':
    args = build_arg_parser().parse_args()
    user = args.user
```

Load the data from the movie ratings file `ratings.json`. This file contains the names of people and their ratings for various movies:

```
ratings_file = 'ratings.json'

with open(ratings_file, 'r') as f:
    data = json.loads(f.read())
```

Find the top three users who are similar to the user specified by the input argument. You can change it to any number of users depending on your choice. Print the output along with the scores:

```
print('\nUsers similar to ' + user + ':\n')
similar_users = find_similar_users(data, user, 3)
print('User\t\t\tSimilarity score')
print('-'*41)
for item in similar_users:
    print(item[0], '\t\t', round(float(item[1]), 2))
```

The full code is given in the file `collaborative_filtering.py`. Let's run the code and find out the users who are similar to Bill Duffy:

```
$ python3 collaborative_filtering.py --user "Bill Duffy"
```

You will get the following output on your Terminal:

```
Users similar to Bill Duffy:

User                    Similarity score
----------------------------------------
David Smith             0.99
Samuel Miller           0.88
Adam Cohen              0.86
```

Let's run the code and find out the users who are similar to `Clarissa Jackson`:

```
$ python3 collaborative_filtering.py --user "Clarissa Jackson"
```

You will get the following output on your Terminal:

```
Users similar to Clarissa Jackson:

User                        Similarity score
----------------------------------------
Chris Duncan                1.0
Bill Duffy                  0.83
Samuel Miller               0.73
```

Building a movie recommendation system

Now that we have all the building blocks in place, it's time to build a movie recommendation system. We learned all the underlying concepts that are needed to build a recommendation system. In this section, we will build a movie recommendation system based on the data provided in the file `ratings.json`. This file contains a set of people and their ratings for various movies. When we want to find movie recommendations for a given user, we will need to find similar users in the dataset and then come up with recommendations for this person.

Create a new Python file and import the following packages:

```
import argparse
import json
import numpy as np

from compute_scores import pearson_score
from collaborative_filtering import find_similar_users
```

Define a function to parse the input arguments. The only input argument would be the name of the user:

```
def build_arg_parser():
    parser = argparse.ArgumentParser(description='Find the movie recommendations for the given user')
    parser.add_argument('--user', dest='user', required=True,
            help='Input user')
    return parser
```

Define a function to get the movie recommendations for a given user. If the user doesn't exist in the dataset, raise an error:

```
# Get movie recommendations for the input user
def get_recommendations(dataset, input_user):
    if input_user not in dataset:
        raise TypeError('Cannot find ' + input_user + ' in the dataset')
```

Define the variables to track the scores:

```
overall_scores = {}
similarity_scores = {}
```

Compute a similarity score between the input user and all the other users in the dataset:

```
for user in [x for x in dataset if x != input_user]:
    similarity_score = pearson_score(dataset, input_user, user)
```

If the similarity score is less than 0, you can continue with the next user in the dataset:

```
if similarity_score <= 0:
    continue
```

Extract a list of movies that have been rated by the current user but haven't been rated by the input user:

```
filtered_list = [x for x in dataset[user] if x not in \
    dataset[input_user] or dataset[input_user][x] == 0]
```

For each item in the filtered list, keep a track of the weighted rating based on the similarity score. Also keep a track of the similarity scores:

```
for item in filtered_list:
    overall_scores.update({item: dataset[user][item] * similarity_score})
    similarity_scores.update({item: similarity_score})
```

If there are no such movies, then we cannot recommend anything:

```
if len(overall_scores) == 0:
    return ['No recommendations possible']
```

Normalize the scores based on the weighted scores:

```
# Generate movie ranks by normalization
movie_scores = np.array([[score/similarity_scores[item], item]
    for item, score in overall_scores.items()])
```

Building Recommender Systems

Sort the scores and extract the movie recommendations:

```
# Sort in decreasing order
movie_scores = movie_scores[np.argsort(movie_scores[:, 0])[::-1]]

# Extract the movie recommendations
movie_recommendations = [movie for _, movie in movie_scores]

return movie_recommendations
```

Define the main function and parse the input arguments to extract the name of the input user:

```
if __name__=='__main__':
    args = build_arg_parser().parse_args()
    user = args.user
```

Load the movie ratings data from the file `ratings.json`:

```
ratings_file = 'ratings.json'

with open(ratings_file, 'r') as f:
    data = json.loads(f.read())
```

Extract the movie recommendations and print the output:

```
print("\nMovie recommendations for " + user + ":")
movies = get_recommendations(data, user)
for i, movie in enumerate(movies):
    print(str(i+1) + '. ' + movie)
```

The full code is given in the file `movie_recommender.py`. Let's find out the movie recommendations for `Chris Duncan`:

```
$ python3 movie_recommender.py --user "Chris Duncan"
```

You will see the following output on your Terminal:

```
Movie recommendations for Chris Duncan:
1. Vertigo
2. Goodfellas
3. Scarface
4. Roman Holiday
```

Let's find out the movie recommendations for `Julie Hammel`:

```
$ python3 movie_recommender.py --user "Julie Hammel"
```

You will see the following output on your Terminal:

```
Movie recommendations for Julie Hammel:
1. The Apartment
2. Vertigo
3. Raging Bull
```

Summary

In this chapter, we learned how to create a data processor pipeline that can be used to train a machine-learning system. We learned how to extract K nearest neighbors to any given data point from a given dataset. We then used this concept to build the K Nearest Neighbors classifier. We discussed how to compute similarity scores such as the Euclidean and Pearson scores. We learned how to use collaborative filtering to find similar users from a given dataset and used it to build a movie recommendation system.

In the next chapter, we will learn about logic programming and see how to build an inference engine that can solve a real world problem.

6
Logic Programming

In this chapter, we are going to learn how to write programs using logic programming. We will discuss various programming paradigms and see how programs are constructed with logic programming. We will learn about the building blocks of logic programming and see how to solve problems in this domain. We will implement Python programs to build various solvers that solve a variety of problems.

By the end of this chapter, you will know about the following:

- What is logic programming?
- Understanding the building blocks of logic programming
- Solving problems using logic programming
- Installing Python packages
- Matching mathematical expressions
- Validating primes
- Parsing a family tree
- Analyzing geography
- Building a puzzle solver

What is logic programming?

Logic programming is a programming paradigm, which basically means it is a particular way to approach programming. Before we talk about what it constitutes and how it is relevant in Artificial Intelligence, let's talk a bit about programming paradigms.

The concept of programming paradigms arises owing to the need to classify programming languages. It refers to the way computer programs solve problems through code. Some programming paradigms are primarily concerned with implications or the sequence of operations used to achieve the result. Other programming paradigms are concerned about how we organize the code.

Here are some of the more popular programming paradigms:

- **Imperative**: This uses statements to change a program's state, thus allowing for side effects.
- **Functional**: This treats computation as an evaluation of mathematical functions and does not allow changing states or mutable data.
- **Declarative**: This is a way of programming where you write your programs by describing what you want to do and not how you want to do it. You express the logic of the underlying computation without explicitly describing the control flow.
- **Object Oriented**: This groups the code within the program in such a way that each object is responsible for itself. The objects contain data and methods that specify how the changes happen.
- **Procedural**: This groups the code into functions and each function is responsible for a particular series of steps.
- **Symbolic**: This uses a particular style of syntax and grammar through which the program can modify its own components by treating them as plain data.
- **Logic**: This views computation as automatic reasoning over a database of knowledge consisting of facts and rules.

In order to understand logic programming, let's understand the concepts of computation and deduction. To compute something, we start with an expression and a set of rules. This set of rules is basically the program.

We use these expressions and rules to generate the output. For example, let's say we want to compute the sum of 23, 12, and 49:

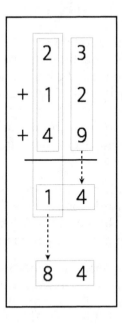

The procedure would be as follows:

23 + 12 + 49 => (2 + 1 + 4 + 1)4 => 84

On the other hand, if we want to deduce something, we need to start from a conjecture. We then need to construct a proof according to a set of rules. In essence, the process computation is mechanical, whereas the process of deduction is more creative.

When we write a program in the logic programming paradigm, we specify a set of statements based on facts and rules about the problem domain and the solver solves it using this information.

Understanding the building blocks of logic programming

In programming object-oriented or imperative paradigms, we always have to specify how a variable is defined. In logic programming, things work a bit differently. We can pass an uninstantiated argument to a function and the interpreter will instantiate these variables for us by looking at the facts defined by the user. This is a powerful way of approaching the variable matching problem. The process of matching variables with different items is called unification. This is one of the places logic programming really stands apart. We need to specify something called relations in logic programming. These relations are defined by means of clauses called facts and rules.

Facts are just statements that are truths about our program and the data that it's operating on. The syntax is pretty straightforward. For example, Donald is Allan's son, can be a fact whereas, Who is Allan's son? cannot be a fact. Every logic program needs facts to work with, so that it can achieve the given goal based on them.

Rules are the things we have learned about how to express various facts and how to query them. They are the constraints that we have to work with and they allow us to make conclusions about the problem domain. For example, let's say you are working on building a chess engine. You need to specify all the rules about how each piece can move on the chessboard. In essence, the final conclusion is valid only if all the relations are true.

Solving problems using logic programming

Logic programming looks for solutions by using facts and rules. We need to specify a goal for each program. In the case where a logic program and a goal don't contain any variables, the solver comes up with a tree that constitutes the search space for solving the problem and getting to the goal.

One of the most important things about logic programming is how we treat the rules. Rules can be viewed as logical statements. Let's consider the following:

Kathy likes chocolate => Alexander loves Kathy

This can be read as an implication that says, *If Kathy likes chocolate, then Alexander loves Kathy*. It can also be construed as *Kathy likes chocolate implies Alexander loves Kathy*. Similarly, let's consider the following rule:

Crime movies, English => Martin Scorsese

It can be read as the implication *If you like crime movies in English, then you would like movies made by Martin Scorsese*.

This construction is used in various forms throughout logic programming to solve various types of problems. Let's go ahead and see how to solve these problems in Python.

Installing Python packages

Before we start logic programming in Python, we need to install a couple of packages. The package `logpy` is a Python package that enables logic programming in Python. We will also be using SymPy for some of the problems. So let's go ahead and install `logpy` and `sympy` using `pip`:

```
$ pip3 install logpy
$ pip3 install sympy
```

If you get an error during the installation process for `logpy`, you can install from source at https://github.com/logpy/logpy. Once you have successfully installed these packages, you can proceed to the next section.

Matching mathematical expressions

We encounter mathematical operations all the time. Logic programming is a very efficient way of comparing expressions and finding out unknown values. Let's see how to do that.

Create a new Python file and import the following packages:

```
from logpy import run, var, fact
import logpy.assoccomm as la
```

Define a couple of mathematical operations:

```
# Define mathematical operations
add = 'addition'
mul = 'multiplication'
```

Logic Programming

Both addition and multiplication are commutative operations. Let's specify that:

```
# Declare that these operations are commutative
# using the facts system
fact(la.commutative, mul)
fact(la.commutative, add)
fact(la.associative, mul)
fact(la.associative, add)
```

Let's define some variables:

```
# Define some variables
a, b, c = var('a'), var('b'), var('c')
```

Consider the following expression:

```
expression_orig = 3 x (-2) + (1 + 2 x 3) x (-1)
```

Let's generate this expression with masked variables. The first expression would be:

- *expression1 = (1 + 2 x a) x b + 3 x c*

The second expression would be:

- *expression2 = c x 3 + b x (2 x a + 1)*

The third expression would be:

- *expression3 = (((2 x a) x b) + b) + 3 x c*

If you observe carefully, all three expressions represent the same basic expression. Our goal is to match these expressions with the original expression to extract the unknown values:

```
# Generate expressions
expression_orig = (add, (mul, 3, -2), (mul, (add, 1, (mul, 2, 3)), -1))
expression1 = (add, (mul, (add, 1, (mul, 2, a)), b), (mul, 3, c))
expression2 = (add, (mul, c, 3), (mul, b, (add, (mul, 2, a), 1)))
expression3 = (add, (add, (mul, (mul, 2, a), b), b), (mul, 3, c))
```

Compare the expressions with the original expression. The method run is commonly used in `logpy`. This method takes the input arguments and runs the expression. The first argument is the number of values, the second argument is a variable, and the third argument is a function:

```
# Compare expressions
print(run(0, (a, b, c), la.eq_assoccomm(expression1, expression_orig)))
print(run(0, (a, b, c), la.eq_assoccomm(expression2, expression_orig)))
print(run(0, (a, b, c), la.eq_assoccomm(expression3, expression_orig)))
```

The full code is given in `expression_matcher.py`. If you run the code, you will see the following output on your Terminal:

```
((3, -1, -2),)
((3, -1, -2),)
()
```

The three values in the first two lines represent the values for a, b, and c. The first two expressions matched with the original expression, whereas the third one returned nothing. This is because even though the third expression is mathematically the same, it is structurally different. Pattern comparison works by comparing the structure of the expressions.

Validating primes

Let's see how to use logic programming to check for prime numbers. We will use the constructs available in `logpy` to determine which numbers in the given list are prime, as well as finding out if a given number is a prime or not.

Create a new Python file and import the following packages:

```
import itertools as it
import logpy.core as lc
from sympy.ntheory.generate import prime, isprime
```

Logic Programming

Next, define a function that checks if the given number is prime depending on the type of data. If it's a number, then it's pretty straightforward. If it's a variable, then we have to run the sequential operation. To give a bit of background, the method `conde` is a goal constructor that provides logical AND and OR operations. The method `condeseq` is like `conde`, but it supports generic iterator of goals:

```
# Check if the elements of x are prime
def check_prime(x):
    if lc.isvar(x):
        return lc.condeseq([(lc.eq, x, p)] for p in map(prime, it.count(1)))
    else:
        return lc.success if isprime(x) else lc.fail
```

Declare the variable x that will be used:

```
# Declate the variable
x = lc.var()
```

Define a set of numbers and check which numbers are prime. The method `membero` checks if a given number is a member of the list of numbers specified in the input argument:

```
# Check if an element in the list is a prime number
list_nums = (23, 4, 27, 17, 13, 10, 21, 29, 3, 32, 11, 19)
print('\nList of primes in the list:')
print(set(lc.run(0, x, (lc.membero, x, list_nums), (check_prime, x))))
```

Let's use the function in a slightly different way now by printing the first 7 prime numbers:

```
# Print first 7 prime numbers
print('\nList of first 7 prime numbers:')
print(lc.run(7, x, check_prime(x)))
```

The full code is given in `prime.py`. If you run the code, you will see the following output:

```
List of primes in the list:
{3, 11, 13, 17, 19, 23, 29}
List of first 7 prime numbers:
(2, 3, 5, 7, 11, 13, 17)
```

You can confirm that the output values are correct.

Parsing a family tree

Now that we are more familiar with logic programming, let's use it to solve an interesting problem. Consider the following family tree:

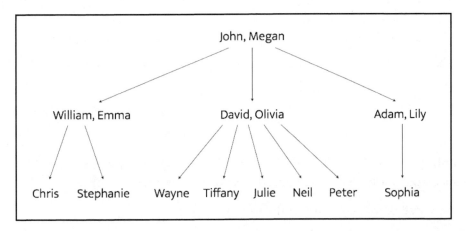

John and Megan have three sons – William, David, and Adam. The wives of William, David, and Adam are Emma, Olivia, and Lily respectively. William and Emma have two children – Chris and Stephanie. David and Olivia have five children – Wayne, Tiffany, Julie, Neil, and Peter. Adam and Lily have one child – Sophia. Based on these facts, we can create a program that can tell us the name of Wayne's grandfather or Sophia's uncles are. Even though we have not explicitly specified anything about the grandparent or uncle relationships, logic programming can infer them.

These relationships are specified in a file called relationships.json provided for you. The file looks like the following:

```
{
    "father":
    [
        {"John": "William"},
        {"John": "David"},
        {"John": "Adam"},
        {"William": "Chris"},
        {"William": "Stephanie"},
        {"David": "Wayne"},
        {"David": "Tiffany"},
        {"David": "Julie"},
        {"David": "Neil"},
        {"David": "Peter"},
        {"Adam": "Sophia"}
```

```
        ],
        "mother":
        [
                {"Megan": "William"},
                {"Megan": "David"},
                {"Megan": "Adam"},
                {"Emma": "Stephanie"},
                {"Emma": "Chris"},
                {"Olivia": "Tiffany"},
                {"Olivia": "Julie"},
                {"Olivia": "Neil"},
                {"Olivia": "Peter"},
                {"Lily": "Sophia"}
        ]
}
```

It is a simple json file that specifies only the father and mother relationships. Note that we haven't specified anything about husband and wife, grandparents, or uncles.

Create a new Python file and import the following packages:

```
import json
from logpy import Relation, facts, run, conde, var, eq
```

Define a function to check if x is the parent of y. We will use the logic that if x is the parent of y, then x is either the father or the mother. We have already defined "father" and "mother" in our fact base:

```
# Check if 'x' is the parent of 'y'
def parent(x, y):
    return conde([father(x, y)], [mother(x, y)])
```

Define a function to check if x is the grandparent of y. We will use the logic that if x is the grandparent of y, then the offspring of x will be the parent of y:

```
# Check if 'x' is the grandparent of 'y'
def grandparent(x, y):
    temp = var()
    return conde((parent(x, temp), parent(temp, y)))
```

Define a function to check if x is the sibling of y. We will use the logic that if x is the sibling of y, then x and y will have the same parents. Notice that there is a slight modification needed here because when we list out all the siblings of x, x will be listed as well because x satisfies these conditions. So when we print the output, we will have to remove x from the list. We will discuss this in the main function:

```
# Check for sibling relationship between 'a' and 'b'
def sibling(x, y):
    temp = var()
    return conde((parent(temp, x), parent(temp, y)))
```

Define a function to check if x is y's uncle. We will use the logic that if x is y's uncle, then x grandparents will be the same as y's parents. Notice that there is a slight modification needed here because when we list out all the uncles of x, x's father will be listed as well because x's father satisfies these conditions. So when we print the output, we will have to remove x's father from the list. We will discuss this in the main function:

```
# Check if x is y's uncle
def uncle(x, y):
    temp = var()
    return conde((father(temp, x), grandparent(temp, y)))
```

Define the main function and initialize the relations `father` and `mother`:

```
if __name__=='__main__':
    father = Relation()
    mother = Relation()
```

Load the data from the `relationships.json` file:

```
    with open('relationships.json') as f:
        d = json.loads(f.read())
```

Read the data and add them to our fact base:

```
    for item in d['father']:
        facts(father, (list(item.keys())[0], list(item.values())[0]))

    for item in d['mother']:
        facts(mother, (list(item.keys())[0], list(item.values())[0]))
```

Define the variable x:

```
x = var()
```

We are now ready to ask some questions and see if our solver can come up with the right answers. Let's ask who John's children are:

```
# John's children
name = 'John'
output = run(0, x, father(name, x))
print("\nList of " + name + "'s children:")
for item in output:
    print(item)
```

Who is William's mother?

```
# William's mother
name = 'William'
output = run(0, x, mother(x, name))[0]
print("\n" + name + "'s mother:\n" + output)
```

Who are Adam's parents?

```
# Adam's parents
name = 'Adam'
output = run(0, x, parent(x, name))
print("\nList of " + name + "'s parents:")
for item in output:
    print(item)
```

Who are Wayne's grandparents?

```
# Wayne's grandparents
name = 'Wayne'
output = run(0, x, grandparent(x, name))
print("\nList of " + name + "'s grandparents:")
for item in output:
    print(item)
```

Who are Megan's grandchildren?

```
# Megan's grandchildren
name = 'Megan'
output = run(0, x, grandparent(name, x))
print("\nList of " + name + "'s grandchildren:")
for item in output:
    print(item)
```

Who are David's siblings?

```
# David's siblings
name = 'David'
output = run(0, x, sibling(x, name))
siblings = [x for x in output if x != name]
print("\nList of " + name + "'s siblings:")
for item in siblings:
    print(item)
```

Who are Tiffany's uncles?

```
# Tiffany's uncles
name = 'Tiffany'
name_father = run(0, x, father(x, name))[0]
output = run(0, x, uncle(x, name))
output = [x for x in output if x != name_father]
print("\nList of " + name + "'s uncles:")
for item in output:
    print(item)
```

List out all the spouses in the family:

```
# All spouses
a, b, c = var(), var(), var()
output = run(0, (a, b), (father, a, c), (mother, b, c))
print("\nList of all spouses:")
for item in output:
    print('Husband:', item[0], '<==> Wife:', item[1])
```

Logic Programming

The full code is given in `family.py`. If you run the code, you will see many things on your Terminal. The first half looks like the following:

```
List of John's children:
David
William
Adam

William's mother:
Megan

List of Adam's parents:
John
Megan

List of Wayne's grandparents:
John
Megan
```

The second half looks like the following:

```
List of Megan's grandchildren:
Chris
Sophia
Peter
Stephanie
Julie
Tiffany
Neil
Wayne

List of David's siblings:
William
Adam

List of Tiffany's uncles:
William
Adam

List of all spouses:
Husband: Adam <==> Wife: Lily
Husband: David <==> Wife: Olivia
Husband: John <==> Wife: Megan
Husband: William <==> Wife: Emma
```

You can compare the outputs with the family tree to ensure that the answers are indeed correct.

Analyzing geography

Let's use logic programming to build a solver to analyze geography. In this problem, we will specify information about the location of various states in the US and then query our program to answer various questions based on those facts and rules. The following is a map of the US:

You have been provided with two text files named `adjacent_states.txt` and `coastal_states.txt`. These files contain the details about which states are adjacent to each other and which states are coastal. Based on this, we can get interesting information like What states are adjacent to both Oklahoma and Texas? or Which coastal state is adjacent to both New Mexico and Louisiana?

Logic Programming

Create a new Python file and import the following:

```
from logpy import run, fact, eq, Relation, var
```

Initialize the relations:

```
adjacent = Relation()
coastal = Relation()
```

Define the input files to load the data from:

```
file_coastal = 'coastal_states.txt'
file_adjacent = 'adjacent_states.txt'
```

Load the data:

```
# Read the file containing the coastal states
with open(file_coastal, 'r') as f:
    line = f.read()
    coastal_states = line.split(',')
```

Add the information to the fact base:

```
# Add the info to the fact base
for state in coastal_states:
    fact(coastal, state)
```

Read the adjacency data:

```
# Read the file containing the coastal states
with open(file_adjacent, 'r') as f:
    adjlist = [line.strip().split(',') for line in f if line and
line[0].isalpha()]
```

Add the adjacency information to the fact base:

```
# Add the info to the fact base
for L in adjlist:
    head, tail = L[0], L[1:]
    for state in tail:
        fact(adjacent, head, state)
```

Initialize the variables x and y:

```
# Initialize the variables
x = var()
y = var()
```

[168]

We are now ready to ask some questions. Check if Nevada is adjacent to Louisiana:

```
# Is Nevada adjacent to Louisiana?
output = run(0, x, adjacent('Nevada', 'Louisiana'))
print('\nIs Nevada adjacent to Louisiana?:')
print('Yes' if len(output) else 'No')
```

Print out all the states that are adjacent to Oregon:

```
# States adjacent to Oregon
output = run(0, x, adjacent('Oregon', x))
print('\nList of states adjacent to Oregon:')
for item in output:
    print(item)
```

List all the coastal states that are adjacent to Mississippi:

```
# States adjacent to Mississippi that are coastal
output = run(0, x, adjacent('Mississippi', x), coastal(x))
print('\nList of coastal states adjacent to Mississippi:')
for item in output:
    print(item)
```

List seven states that border a coastal state:

```
# List of 'n' states that border a coastal state
n = 7
output = run(n, x, coastal(y), adjacent(x, y))
print('\nList of ' + str(n) + ' states that border a coastal state:')
for item in output:
    print(item)
```

List states that are adjacent to both Arkansas and Kentucky:

```
# List of states that adjacent to the two given states
output = run(0, x, adjacent('Arkansas', x), adjacent('Kentucky', x))
print('\nList of states that are adjacent to Arkansas and Kentucky:')
for item in output:
    print(item)
```

Logic Programming

The full code is given in `states.py`. If you run the code, you will see the following output:

```
Is Nevada adjacent to Louisiana?:
No

List of states adjacent to Oregon:
Washington
California
Nevada
Idaho

List of coastal states adjacent to Mississippi:
Alabama
Louisiana

List of 7 states that border a coastal state:
Georgia
Pennsylvania
Massachusetts
Wisconsin
Maine
Oregon
Ohio

List of states that are adjacent to Arkansas and Kentucky:
Missouri
Tennessee
```

You can cross-check the output with the US map to verify if the answers are right. You can also add more questions to the program to see if it can answer them.

Building a puzzle solver

Another interesting application of logic programming is in solving puzzles. We can specify the conditions of a puzzle and the program will come up with a solution. In this section, we will specify various bits and pieces of information about four people and ask for the missing piece of information.

In the logic program, we specify the puzzle as follows:

- Steve has a blue car
- The person who owns the cat lives in Canada
- Matthew lives in USA
- The person with the black car lives in Australia
- Jack has a cat
- Alfred lives in Australia
- The person who has a dog lives in France
- Who has a rabbit?

The goal is the find the person who has a rabbit. Here are the full details about the four people:

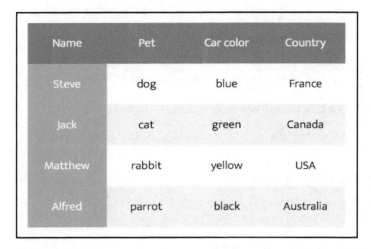

Create a new Python file and import the following packages:

```
from logpy import *
from logpy.core import lall
```

Declare the variable `people`:

```
# Declare the variable
people = var()
```

Define all the rules using `lall`. The first rule is that there are four people:

```
# Define the rules
rules = lall(
    # There are 4 people
    (eq, (var(), var(), var(), var()), people),
```

The person named Steve has a blue car:

```
    # Steve's car is blue
    (membero, ('Steve', var(), 'blue', var()), people),
```

Logic Programming

The person who has a cat lives in Canada:

```
# Person who has a cat lives in Canada
(membero, (var(), 'cat', var(), 'Canada'), people),
```

The person named Matthew lives in USA:

```
# Matthew lives in USA
(membero, ('Matthew', var(), var(), 'USA'), people),
```

The person who has a black car lives in Australia:

```
# The person who has a black car lives in Australia
(membero, (var(), var(), 'black', 'Australia'), people),
```

The person named Jack has a cat:

```
# Jack has a cat
(membero, ('Jack', 'cat', var(), var()), people),
```

The person named Alfred lives in Australia:

```
# Alfred lives in Australia
(membero, ('Alfred', var(), var(), 'Australia'), people),
```

The person who has a dog lives in France:

```
# Person who owns the dog lives in France
(membero, (var(), 'dog', var(), 'France'), people),
```

One of the people in this group has a rabbit. Who is that person?

```
# Who has a rabbit?
(membero, (var(), 'rabbit', var(), var()), people)
)
```

Run the solver with the preceding constraints:

```
# Run the solver
solutions = run(0, people, rules)
```

Extract the output from the solution:

```
# Extract the output
output = [house for house in solutions[0] if 'rabbit' in house][0][0]
```

[172]

Print the full matrix obtained from the solver:

```
# Print the output
print('\n' + output + ' is the owner of the rabbit')
print('\nHere are all the details:')
attribs = ['Name', 'Pet', 'Color', 'Country']
print('\n' + '\t\t'.join(attribs))
print('=' * 57)
for item in solutions[0]:
    print('')
    print('\t\t'.join([str(x) for x in item]))
```

The full code is given in `puzzle.py`. If you run the code, you will see the following output:

```
Matthew is the owner of the rabbit

Here are all the details:

Name            Pet             Color           Country
=========================================================

Steve           dog             blue            France

Jack            cat             ~_9             Canada

Matthew         rabbit          ~_11            USA

Alfred          ~_13            black           Australia
```

The preceding figure shows all the values obtained using the solver. Some of them are still unknown as indicated by numbered names. Even though the information was incomplete, our solver was able to answer our question. But in order to answer every single question, you may need to add more rules. This program was to demonstrate how to solve a puzzle with incomplete information. You can play around with it and see how you can build puzzle solvers for various scenarios.

Summary

In this chapter, we learned how to write Python programs using logic programming. We discussed how various programming paradigms deal with building programs. We understood how programs are built in logic programming. We learned about various building blocks of logic programming and discussed how to solve problems in this domain.

We implemented various Python programs to solve interesting problems and puzzles. In the next chapter, we will learn about heuristic search techniques and use those algorithms to solve real world problems.

7
Heuristic Search Techniques

In this chapter, we are going to learn about heuristic search techniques. Heuristic search techniques are used to search through the solution space to come up with answers. The search is conducted using heuristics that guide the search algorithm. This heuristic allows the algorithm to speed up the process, which would otherwise take a really long time to arrive at the solution.

By the end of this chapter, you will know about the following:

- What is heuristic search?
- Uninformed vs. informed search
- Constraint Satisfaction Problems
- Local search techniques
- Simulated annealing
- Constructing a string using greedy search
- Solving a problem with constraints
- Solving the region coloring problem
- Building an 8-puzzle solver
- Building a maze solver

What is heuristic search?

Searching and organizing data is an important topic within Artificial Intelligence. There are many problems that require searching for an answer within the solution domain. There are many possible solutions to a given problem and we do not know which ones are correct. By efficiently organizing the data, we can search for solutions quickly and effectively.

More often, there are so many possible options to solve a given problem that no algorithm can be developed to find a right solution. Also, going through every single solution is not possible because it is prohibitively expensive. In such cases, we rely on a rule of thumb that helps us narrow down the search by eliminating the options that are obviously wrong. This rule of thumb is called a **heuristic**. The method of using heuristics to guide our search is called **heuristic search**.

Heuristic techniques are very handy because they help us speed up the process. Even if the heuristic is not able to completely eliminate some options, it will help us to order those options so that we are more likely to get to the better solutions first.

Uninformed versus Informed search

If you are familiar with computer science, you should have heard about search techniques like **Depth First Search (DFS)**, **Breadth First Search (BFS)**, and **Uniform Cost Search (UCS)**. These are search techniques that are commonly used on graphs to get to the solution. These are examples of uninformed search. They do not use any prior information or rules to eliminate some paths. They check all the plausible paths and pick the optimal one.

Heuristic search, on the other hand, is called **Informed search** because it uses prior information or rules to eliminate unnecessary paths. Uninformed search techniques do not take the goal into account. These techniques don't really know where they are trying to go unless they just stumble upon the goal in the process.

In the graph problem, we can use heuristics to guide the search. For example, at each node, we can define a heuristic function that returns a score that represents the estimate of the cost of the path from the current node to the goal. By defining this heuristic function, we are informing the search technique about the right direction to reach the goal. This will allow the algorithm to identify which neighbor will lead to the goal.

We need to note that heuristic search might not always find the most optimal solution. This is because we are not exploring every single possibility and we are relying on a heuristic. But it is guaranteed to find a good solution in a reasonable time, which is what we expect from a practical solution. In real-world scenarios, we need solutions that are fast and effective. Heuristic searches provide an efficient solution by arriving at a reasonable solution quickly. They are used in cases where the problems cannot be solved in any other way or would take a really long time to solve.

Constraint Satisfaction Problems

There are many problems that have to be solved under constraints. These constraints are basically conditions that cannot be violated during the process of solving the problem. These problems are referred to as **Constraint Satisfaction Problems** (**CSPs**).

CSPs are basically mathematical problems that are defined as a set of variables that must satisfy a number of constraints. When we arrive at the final solution, the states of the variables must obey all the constraints. This technique represents the entities involved in a given problem as a collection of a fixed number of constraints over variables. These variables need to be solved by constraint satisfaction methods.

These problems require a combination of heuristics and other search techniques to be solved in a reasonable amount of time. In this case, we will use constraint satisfaction techniques to solve problems on finite domains. A finite domain consists of a finite number of elements. Since we are dealing with finite domains, we can use search techniques to arrive at the solution.

Local search techniques

Local search is a particular way of solving a CSP. It keeps improving the values until all the constraints are satisfied. It iteratively keeps updating the variables until we arrive at the destination. These algorithms modify the value during each step of the process that gets us closer to the goal. In the solution space, the updated value is closer to the goal than the previous value. Hence it is known as a local search.

Local search algorithm is a heuristic search algorithm. These algorithms use a function that calculates the quality of each update. For example, it can count the number of constraints that are being violated by the current update or it can see how the update affects the distance to the goal. This is referred to as the cost of the assignment. The overall goal of local search is to find the minimal cost update at each step.

Hill climbing is a popular local search technique. It uses a heuristic function that measures the difference between the current state and the goal. When we start, it checks if the state is the final goal. If it is, then it stops. If not, then it selects an update and generates a new state. If it's closer to the goal than the current state, then it makes that the current state. If not, it ignores it and continues the process until it checks all possible updates. It basically climbs the hill until it reaches the summit.

Simulated Annealing

Simulated Annealing is a type of local search as well as a stochastic search technique. Stochastic search techniques are used extensively in various fields such as robotics, chemistry, manufacturing, medicine, economics, and so on. We can perform things like optimizing the design of a robot, determining the timing strategies for automated control in factories, and planning traffic. Stochastic algorithms are used to solve many real-world problems.

Simulated Annealing is a variation of the hill climbing technique. One of the main problems of hill climbing is that it ends up climbing false foothills. This means that it gets stuck in local maxima. So it is better to check out the whole space before we make any climbing decisions. In order to achieve this, the whole space is initially explored to see what it is like. This helps us avoid getting stuck in a plateau or local maxima.

In Simulated Annealing, we reformulate the problem and solve it for minimization, as opposed to maximization. So, we are now descending into valleys as opposed to climbing hills. We are pretty much doing the same thing, but in a different way. We use an objective function to guide the search. This objective function serves as our heuristic.

The reason it is called Simulated Annealing is because it is derived from the metallurgical process. We first heat metals up and then let them cool until they reach the optimal energy state.

The rate at which we cool the system is called the **annealing schedule**. The rate of cooling is important because it directly impacts the final result. In the real world case of metals, if the rate of cooling is too fast, it ends up settling for the local maximum. For example, if we take the heated metal and put it in cold water, it ends up quickly settling for the sub-optimal local maximum.

If the rate of cooling is slow and controlled, we give the metal a chance to arrive at the globally optimum state. The chances of taking big steps quickly towards any particular hill are lower in this case. Since the rate of cooling is slow, it will take its time to choose the best state. We do something similar with data in our case.

We first evaluate the current state and see if it is the goal. If it is, then we stop. If not, then we set the best state variable to the current state. We then define our annealing schedule that controls how quickly it descends into a valley. We compute the difference between the current state and the new state. If the new state is not better, then we make it the current state with a certain predefined probability. We do this using a random number generator and making a decision based on a threshold. If it is above the threshold, then we set the best state to this state. Based on this, we update the annealing schedule depending on the number of nodes. We keep doing this until we arrive at the goal.

Constructing a string using greedy search

Greedy search is an algorithmic paradigm that makes the locally optimal choice at each stage in order to find the global optimum. But in many problems, greedy algorithms do not produce globally optimum solutions. An advantage of using greedy algorithms is that they produce an approximate solution in a reasonable time. The hope is that this approximate solution is reasonably close to the global optimal solution.

Greedy algorithms do not refine their solutions based on new information during the search. For example, let's say you are planning on a road trip and you want to take the best route possible. If you use a greedy algorithm to plan the route, it would ask you to take routes that are shorter but might end up taking more time. It can also lead you to paths that may seem faster in the short term, but might lead to traffic jams later. This happens because greedy algorithms only think about the next step and not the globally optimal final solution.

Let's see how to solve a problem using a greedy search. In this problem, we will try to recreate the input string based on the alphabets. We will ask the algorithm to search the solution space and construct a path to the solution.

We will be using a package called `simpleai` throughout this chapter. It contains various routines that are useful in building solutions using heuristic search techniques. It's available at `https://github.com/simpleai-team/simpleai`. We need to make a few changes to the source code in order to make it work in Python3. A file called `simpleai.zip` has been provided along with the code for the book. Unzip this file into a folder called `simpleai`. This folder contains all the necessary changes to the original library necessary to make it work in Python3. Place the `simpleai` folder in the same folder as your code and you'll be able to run your code smoothly.

Create a new Python file and import the following packages:

```
import argparse
import simpleai.search as ss
```

Define a function to parse the input arguments:

```
def build_arg_parser():
    parser = argparse.ArgumentParser(description='Creates the input string \
            using the greedy algorithm')
    parser.add_argument("--input-string", dest="input_string", required=True,
            help="Input string")
    parser.add_argument("--initial-state", dest="initial_state", required=False,
            default='', help="Starting point for the search")
    return parser
```

Create a class that contains the methods needed to solve the problem. This class inherits the `SearchProblem` class available in the library. We just need to override a couple of methods to suit our needs. The first method `set_target` is a custom method that we define to set the target string:

```
class CustomProblem(ss.SearchProblem):
    def set_target(self, target_string):
        self.target_string = target_string
```

The `actions` is a method that comes with a `SearchProblem` and we need to override it. It's responsible for taking the right steps towards the goal. If the length of the current string is less than the length of the target string, it will return the list of possible alphabets to choose from. If not, it will return an empty string:

```
    # Check the current state and take the right action
    def actions(self, cur_state):
        if len(cur_state) < len(self.target_string):
            alphabets = 'abcdefghijklmnopqrstuvwxyz'
            return list(alphabets + ' ' + alphabets.upper())
        else:
            return []
```

a method to compute the result by concatenating the current string and the action that needs to be taken. This method comes with a `SearchProblem` and we are overriding it:

```
    # Concatenate state and action to get the result
    def result(self, cur_state, action):
        return cur_state + action
```

The method `is_goal` is a part of the `SearchProblem` and it's used to check if we have arrived at the goal:

```
# Check if goal has been achieved
def is_goal(self, cur_state):
    return cur_state == self.target_string
```

The method `heuristic` is also a part of the `SearchProblem` and we need to override it. We will define our own heuristic that will be used to solve the problem. We will calculate how far we are from the goal and use that as the heuristic to guide it towards the goal:

```
# Define the heuristic that will be used
    def heuristic(self, cur_state):
        # Compare current string with target string
        dist = sum([1 if cur_state[i] != self.target_string[i] else 0
                  for i in range(len(cur_state))])

        # Difference between the lengths
        diff = len(self.target_string) - len(cur_state)

        return dist + diff
```

e input arguments:

```
if __name__=='__main__':
    args = build_arg_parser().parse_args()
```

Initialize the `CustomProblem` object:

```
# Initialize the object
problem = CustomProblem()
```

Set the starting point as well as the goal we want to achieve:

```
# Set target string and initial state
problem.set_target(args.input_string)
problem.initial_state = args.initial_state
```

Run the solver:

```
# Solve the problem
output = ss.greedy(problem)
```

Print the path to the solution:

```
print('\nTarget string:', args.input_string)
print('\nPath to the solution:')
for item in output.path():
    print(item)
```

The full code is given in the file `greedy_search.py`. If you run the code with an empty initial state:

```
$ python3 greedy_search.py --input-string 'Artificial Intelligence' --initial-state ''
```

You will get the following output:

```
Path to the solution:
(None, '')
('A', 'A')
('r', 'Ar')
('t', 'Art')
('i', 'Arti')
('f', 'Artif')
('i', 'Artifi')
('c', 'Artific')
('i', 'Artifici')
('a', 'Artificia')
('l', 'Artificial')
(' ', 'Artificial ')
('I', 'Artificial I')
('n', 'Artificial In')
('t', 'Artificial Int')
('e', 'Artificial Inte')
('l', 'Artificial Intel')
('l', 'Artificial Intell')
('i', 'Artificial Intelli')
('g', 'Artificial Intellig')
('e', 'Artificial Intellige')
('n', 'Artificial Intelligen')
('c', 'Artificial Intelligenc')
('e', 'Artificial Intelligence')
```

If you run the code with a non-empty starting point:

```
$ python3 greedy_search.py --input-string 'Artificial Intelligence with Python' --initial-state 'Artificial Inte'
```

You will get the following output:

```
Path to the solution:
(None, 'Artificial Inte')
('l', 'Artificial Intel')
('l', 'Artificial Intell')
('i', 'Artificial Intelli')
('g', 'Artificial Intellig')
('e', 'Artificial Intellige')
('n', 'Artificial Intelligen')
('c', 'Artificial Intelligenc')
('e', 'Artificial Intelligence')
(' ', 'Artificial Intelligence ')
('w', 'Artificial Intelligence w')
('i', 'Artificial Intelligence wi')
('t', 'Artificial Intelligence wit')
('h', 'Artificial Intelligence with')
(' ', 'Artificial Intelligence with ')
('P', 'Artificial Intelligence with P')
('y', 'Artificial Intelligence with Py')
('t', 'Artificial Intelligence with Pyt')
('h', 'Artificial Intelligence with Pyth')
('o', 'Artificial Intelligence with Pytho')
('n', 'Artificial Intelligence with Python')
```

Solving a problem with constraints

We have already discussed how Constraint Satisfaction Problems are formulated. Let's apply them to a real-world problem. In this problem, we have a list of names and each name can only take a fixed set of values. We also have a set of constraints between these people that needs to be satisfied. Let's see how to do it.

Create a new Python file and import the following packages:

```
from simpleai.search import CspProblem, backtrack, \
    min_conflicts, MOST_CONSTRAINED_VARIABLE, \
    HIGHEST_DEGREE_VARIABLE, LEAST_CONSTRAINING_VALUE
```

Define the constraint that specifies that all the variables in the input list should have unique values:

```python
# Constraint that expects all the different variables
# to have different values
def constraint_unique(variables, values):
    # Check if all the values are unique
    return len(values) == len(set(values))
```

Define the constraint that specifies that the first variable should be bigger than the second variable:

```python
# Constraint that specifies that one variable
# should be bigger than other
def constraint_bigger(variables, values):
    return values[0] > values[1]
```

Define the constraint that specifies that if the first variable is odd, then the second variable should be even and vice versa:

```python
# Constraint that specifies that there should be
# one odd and one even variables in the two variables
def constraint_odd_even(variables, values):
    # If first variable is even, then second should
    # be odd and vice versa
    if values[0] % 2 == 0:
        return values[1] % 2 == 1
    else:
        return values[1] % 2 == 0
```

Define the `main` function and define the variables:

```python
if __name__=='__main__':
    variables = ('John', 'Anna', 'Tom', 'Patricia')
```

Define the list of values that each variable can take:

```python
    domains = {
        'John': [1, 2, 3],
        'Anna': [1, 3],
        'Tom': [2, 4],
        'Patricia': [2, 3, 4],
    }
```

Define the constraints for various scenarios. In this case, we specify three constraints as follows:

- John, Anna, and Tom should have different values
- Tom's value should be bigger than Anna's value
- If John's value is odd, then Patricia's value should be even and vice versa

Use the following code:

```
constraints = [
    (('John', 'Anna', 'Tom'), constraint_unique),
    (('Tom', 'Anna'), constraint_bigger),
    (('John', 'Patricia'), constraint_odd_even),
]
```

Use the preceding variables and the constraints to initialize the `CspProblem` object:

```
problem = CspProblem(variables, domains, constraints)
```

Compute the solution and print it:

```
print('\nSolutions:\n\nNormal:', backtrack(problem))
```

Compute the solution using the `MOST_CONSTRAINED_VARIABLE` heuristic:

```
print('\nMost constrained variable:', backtrack(problem,
        variable_heuristic=MOST_CONSTRAINED_VARIABLE))
```

Compute the solution using the `HIGHEST_DEGREE_VARIABLE` heuristic:

```
print('\nHighest degree variable:', backtrack(problem,
        variable_heuristic=HIGHEST_DEGREE_VARIABLE))
```

Compute the solution using the `LEAST_CONSTRAINING_VALUE` heuristic:

```
print('\nLeast constraining value:', backtrack(problem,
        value_heuristic=LEAST_CONSTRAINING_VALUE))
```

Compute the solution using the `MOST_CONSTRAINED_VARIABLE` variable heuristic and `LEAST_CONSTRAINING_VALUE` value heuristic:

```
print('\nMost constrained variable and least constraining value:',
        backtrack(problem,
    variable_heuristic=MOST_CONSTRAINED_VARIABLE,
        value_heuristic=LEAST_CONSTRAINING_VALUE))
```

Compute the solution using the `HIGHEST_DEGREE_VARIABLE` variable heuristic and `LEAST_CONSTRAINING_VALUE` value heuristic:

```
print('\nHighest degree and least constraining value:',
        backtrack(problem, variable_heuristic=HIGHEST_DEGREE_VARIABLE,
        value_heuristic=LEAST_CONSTRAINING_VALUE))
```

Compute the solution using the minimum conflicts heuristic:

```
print('\nMinimum conflicts:', min_conflicts(problem))
```

The full code is given in the file `constrained_problem.py`. If you run the code, you will get the following output:

```
Solutions:
Normal: {'Patricia': 2, 'John': 1, 'Anna': 3, 'Tom': 4}
Most constrained variable: {'Patricia': 2, 'John': 3, 'Anna': 1, 'Tom': 2}
Highest degree variable: {'Patricia': 2, 'John': 1, 'Anna': 3, 'Tom': 4}
Least constraining value: {'Patricia': 2, 'John': 1, 'Anna': 3, 'Tom': 4}
Most constrained variable and least constraining value: {'Patricia': 2, 'John': 3, 'Anna': 1, 'Tom': 2}
Highest degree and least constraining value: {'Patricia': 2, 'John': 1, 'Anna': 3, 'Tom': 4}
Minimum conflicts: {'Patricia': 4, 'John': 1, 'Anna': 3, 'Tom': 4}
```

You can check the constraints to see if the solutions satisfy all those constraints.

Solving the region-coloring problem

Let's use the Constraint Satisfaction framework to solve the region-coloring problem. Consider the following screenshot:

Chapter 7

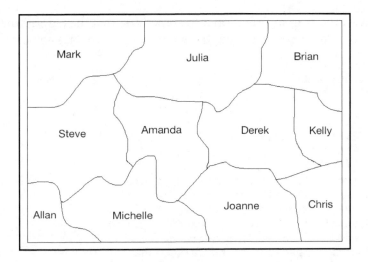

We have a few regions in the preceding figure that are labeled with names. Our goal is to color with four colors so that no adjacent regions have the same color.

Create a new Python file and import the following packages:

```
from simpleai.search import CspProblem, backtrack
```

Define the constraint that specifies that the values should be different:

```
# Define the function that imposes the constraint
# that neighbors should be different
def constraint_func(names, values):
    return values[0] != values[1]
```

Define the `main` function and specify the list of names:

```
if __name__=='__main__':
    # Specify the variables
    names = ('Mark', 'Julia', 'Steve', 'Amanda', 'Brian',
            'Joanne', 'Derek', 'Allan', 'Michelle', 'Kelly')
```

Define the list of possible colors:

```
    # Define the possible colors
    colors = dict((name, ['red', 'green', 'blue', 'gray']) for name in names)
```

[187]

We need to convert the map information into something that the algorithm can understand. Let's define the constraints by specifying the list of people who are adjacent to each other:

```
# Define the constraints
constraints = [
    (('Mark', 'Julia'), constraint_func),
    (('Mark', 'Steve'), constraint_func),
    (('Julia', 'Steve'), constraint_func),
    (('Julia', 'Amanda'), constraint_func),
    (('Julia', 'Derek'), constraint_func),
    (('Julia', 'Brian'), constraint_func),
    (('Steve', 'Amanda'), constraint_func),
    (('Steve', 'Allan'), constraint_func),
    (('Steve', 'Michelle'), constraint_func),
    (('Amanda', 'Michelle'), constraint_func),
    (('Amanda', 'Joanne'), constraint_func),
    (('Amanda', 'Derek'), constraint_func),
    (('Brian', 'Derek'), constraint_func),
    (('Brian', 'Kelly'), constraint_func),
    (('Joanne', 'Michelle'), constraint_func),
    (('Joanne', 'Amanda'), constraint_func),
    (('Joanne', 'Derek'), constraint_func),
    (('Joanne', 'Kelly'), constraint_func),
    (('Derek', 'Kelly'), constraint_func),
]
```

Use the variables and constraints to initialize the object:

```
# Solve the problem
problem = CspProblem(names, colors, constraints)
```

Solve the problem and print the solution:

```
# Print the solution
output = backtrack(problem)
print('\nColor mapping:\n')
for k, v in output.items():
    print(k, '==>', v)
```

The full code is given in the file `coloring.py`. If you run the code, you will get the following output on your Terminal:

If you color the regions based on this output, you will get the following:

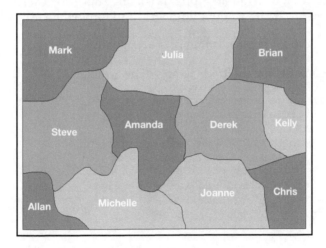

You can check that no two adjacent regions have the same color.

Building an 8-puzzle solver

8-puzzle is a variant of the 15-puzzle. You can check it out at `https://en.wikipedia.org/wiki/15_puzzle`. You will be presented with a randomized grid and your goal is to get it back to the original ordered configuration. You can play the game to get familiar with it at `http://mypuzzle.org/sliding`.

We will use an **A* algorithm** to solve this problem. It is an algorithm that's used to find paths to the solution in a graph. This algorithm is a combination of **Dijkstra's algorithm** and a greedy best-first search. Instead of blindly guessing where to go next, the A* algorithm picks the one that looks the most promising. At each node, we generate the list of all possibilities and then pick the one with the minimal cost required to reach the goal.

Let's see how to define the cost function. At each node, we need to compute the cost. This cost is basically the sum of two costs – the first cost is the cost of getting to the current node and the second cost is the cost of reaching the goal from the current node.

We use this summation as our heuristic. As we can see, the second cost is basically an estimate that's not perfect. If this is perfect, then the A* algorithm arrives at the solution quickly. But it's not usually the case. It takes some time to find the best path to the solution. But A* is very effective in finding the optimal paths and is one of the most popular techniques out there.

Let's use the A* algorithm to build an 8-puzzle solver. This is a variant of the solution given in the `simpleai` library. Create a new Python file and import the following packages:

```
from simpleai.search import astar, SearchProblem
```

Define a class that contains the methods to solve the 8-puzzle:

```
# Class containing methods to solve the puzzle
class PuzzleSolver(SearchProblem):
```

Override the `actions` method to align it with our problem:

```
    # Action method to get the list of the possible
    # numbers that can be moved in to the empty space
    def actions(self, cur_state):
        rows = string_to_list(cur_state)
        row_empty, col_empty = get_location(rows, 'e')
```

Check the location of the empty space and create the new action:

```
        actions = []
        if row_empty > 0:
            actions.append(rows[row_empty - 1][col_empty])
        if row_empty < 2:
            actions.append(rows[row_empty + 1][col_empty])
        if col_empty > 0:
            actions.append(rows[row_empty][col_empty - 1])
        if col_empty < 2:
            actions.append(rows[row_empty][col_empty + 1])

        return actions
```

Override the `result` method. Convert the string to a list and extract the location of the empty space. Generate the result by updating the locations:

```
# Return the resulting state after moving a piece to the empty space
def result(self, state, action):
    rows = string_to_list(state)
    row_empty, col_empty = get_location(rows, 'e')
    row_new, col_new = get_location(rows, action)

    rows[row_empty][col_empty], rows[row_new][col_new] = \
            rows[row_new][col_new], rows[row_empty][col_empty]

    return list_to_string(rows)
```

Check if the goal has been reached:

```
# Returns true if a state is the goal state
def is_goal(self, state):
    return state == GOAL
```

Define the `heuristic` method. We will use the heuristic that computes the distance between the current state and goal state using Manhattan distance:

```
# Returns an estimate of the distance from a state to
# the goal using the manhattan distance
def heuristic(self, state):
    rows = string_to_list(state)

    distance = 0
```

Compute the distance:

```
        for number in '12345678e':
            row_new, col_new = get_location(rows, number)
            row_new_goal, col_new_goal = goal_positions[number]

            distance += abs(row_new - row_new_goal) + abs(col_new - col_new_goal)

        return distance
```

Define a function to convert a list to string:

```
# Convert list to string
def list_to_string(input_list):
    return '\n'.join(['-'.join(x) for x in        input_list])
```

Define a function to convert a string to a list:

[191]

```
# Convert string to list
def string_to_list(input_string):
    return [x.split('-') for x in
        input_string.split('\n')]
```

Define a function to get the location of a given element in the grid:

```
# Find the 2D location of the input element
def get_location(rows, input_element):
    for i, row in enumerate(rows):
        for j, item in enumerate(row):
            if item == input_element:
                return i, j
```

Define the initial state and the final goal we want to achieve:

```
# Final result that we want to achieve
GOAL = '''1-2-3
4-5-6
7-8-e'''

# Starting point
INITIAL = '''1-e-2
6-3-4
7-5-8'''
```

Track the goal positions for each piece by creating a variable:

```
# Create a cache for the goal position of each piece
goal_positions = {}
rows_goal = string_to_list(GOAL)
for number in '12345678e':
    goal_positions[number] = get_location(rows_goal, number)
```

Create the A* solver object using the initial state we defined earlier and extract the result:

```
# Create the solver object
result = astar(PuzzleSolver(INITIAL))
```

Print the solution:

```
# Print the results
for i, (action, state) in enumerate(result.path()):
    print()
    if action == None:
        print('Initial configuration')
    elif i == len(result.path()) - 1:
        print('After moving', action, 'into the empty space. Goal achieved!')
    else:
        print('After moving', action, 'into the empty space')

    print(state)
```

The full code is given in the file `puzzle.py`. If you run the code, you will get a long output on your Terminal. It will start as follows:

```
Initial configuration
1-e-2
6-3-4
7-5-8

After moving 2 into the empty space
1-2-e
6-3-4
7-5-8

After moving 4 into the empty space
1-2-4
6-3-e
7-5-8

After moving 3 into the empty space
1-2-4
6-e-3
7-5-8

After moving 6 into the empty space
1-2-4
e-6-3
7-5-8
```

If you scroll down, you will see the steps taken to arrive at the solution. At the end, you will see the following on your Terminal:

```
After moving 2 into the empty space
e-2-3
1-4-6
7-5-8

After moving 1 into the empty space
1-2-3
e-4-6
7-5-8

After moving 4 into the empty space
1-2-3
4-e-6
7-5-8

After moving 5 into the empty space
1-2-3
4-5-6
7-e-8

After moving 8 into the empty space. Goal achieved!
1-2-3
4-5-6
7-8-e
```

Building a maze solver

Let's use the A* algorithm to solve a maze. Consider the following figure:

```
################################
#        #              #    #
# ####   ########       #    #
# o #    #              #    #
#    ###      #####  ######  #
#        #    ###   #        #
#        #          #        #
#        #    #  #  #  #   ###
#   #####     #     #  # x   #
#             #        #     #
################################
```

The # symbols indicate obstacles. The symbol o represents the starting point and x represents the goal. Our goal is to find the shortest path from the start to the end point. Let's see how to do it in Python. The following solution is a variant of the solution provided in the `simpleai` library. Create a new Python file and import the following packages:

```python
import math
from simpleai.search import SearchProblem, astar
```

Create a class that contains the methods needed to solve the problem:

```python
# Class containing the methods to solve the maze
class MazeSolver(SearchProblem):
```

Define the initializer method:

```python
    # Initialize the class
    def __init__(self, board):
        self.board = board
        self.goal = (0, 0)
```

Extract the initial and final positions:

```python
        for y in range(len(self.board)):
            for x in range(len(self.board[y])):
                if self.board[y][x].lower() == "o":
                    self.initial = (x, y)
                elif self.board[y][x].lower() == "x":
                    self.goal = (x, y)

        super(MazeSolver, self).__init__(initial_state=self.initial)
```

Override the `actions` method. At each position, we need to check the cost of going to the neighboring cells and then append all the possible actions. If the neighboring cell is blocked, then that action is not considered:

```python
    # Define the method that takes actions
    # to arrive at the solution
    def actions(self, state):
        actions = []
        for action in COSTS.keys():
            newx, newy = self.result(state, action)
            if self.board[newy][newx] != "#":
                actions.append(action)

        return actions
```

Override the `result` method. Depending on the current state and the input action, update the x and y coordinates:

```
# Update the state based on the action
def result(self, state, action):
    x, y = state

    if action.count("up"):
        y -= 1
    if action.count("down"):
        y += 1
    if action.count("left"):
        x -= 1
    if action.count("right"):
        x += 1

    new_state = (x, y)

    return new_state
```

Check if we have arrived at the goal:

```
# Check if we have reached the goal
def is_goal(self, state):
    return state == self.goal
```

We need to define the `cost` function. This is the cost of moving to a neighboring cell, and it's different for vertical/horizontal and diagonal moves. We will define these later:

```
# Compute the cost of taking an action
def cost(self, state, action, state2):
    return COSTS[action]
```

Define the heuristic that will be used. In this case, we will use the Euclidean distance:

```
# Heuristic that we use to arrive at the solution
def heuristic(self, state):
    x, y = state
    gx, gy = self.goal

    return math.sqrt((x - gx) ** 2 + (y - gy) ** 2)
```

Define the `main` function and also define the map we discussed earlier:

```
if __name__ == "__main__":
    # Define the map
    MAP = """
    ###############################
    #       #              #  #
    # ####   ########      #  #
    #  o #   #             #  #
    #    ###    #####  ######  #
    #      #  ###   #           #
    #      #    #  # # #      ###
    #    #####    #    # # x   #
    #           #      #       #
    ###############################
    """
```

Convert the map information into a list:

```
    # Convert map to a list
    print(MAP)
    MAP = [list(x) for x in MAP.split("\n") if x]
```

Define the cost of moving around the map. The diagonal move is more expensive than horizontal or vertical moves:

```
    # Define cost of moving around the map
    cost_regular = 1.0
    cost_diagonal = 1.7
```

Assign the costs to the corresponding moves:

```
    # Create the cost dictionary
    COSTS = {
        "up": cost_regular,
        "down": cost_regular,
        "left": cost_regular,
        "right": cost_regular,
        "up left": cost_diagonal,
        "up right": cost_diagonal,
        "down left": cost_diagonal,
        "down right": cost_diagonal,
    }
```

Heuristic Search Techniques

Create a solver object using the custom class we defined earlier:

```
# Create maze solver object
problem = MazeSolver(MAP)
```

Run the solver on the map and extract the result:

```
# Run the solver
result = astar(problem, graph_search=True)
```

Extract the path from the result:

```
# Extract the path
path = [x[1] for x in result.path()]
```

Print the output:

```
# Print the result
print()
for y in range(len(MAP)):
    for x in range(len(MAP[y])):
        if (x, y) == problem.initial:
            print('o', end='')
        elif (x, y) == problem.goal:
            print('x', end='')
        elif (x, y) in path:
            print(' ', end='')
        else:
            print(MAP[y][x], end='')

    print()
```

The full code is given in the file `maze.py`. If you run the code, you will get the following output:

```
###################################
#                 #            #   #
# ####    ########             #   #
#   o #       #                #   #
#     ·###         #####   ######  #
#      · #       ###    #   ····   #
#      · #      #  ·· # ·# #·    ###
#      ·#####    ·#  ·· #   # x    #
#      ········ #           #      #
###################################
```

Summary

In this chapter, we learned how heuristic search techniques work. We discussed the difference between uninformed and informed search. We learned about constraint satisfaction problems and how we can solve problems using this paradigm. We discussed how local search techniques work and why simulated annealing is used in practice. We implemented greedy search for a string problem. We solved a problem using the CSP formulation.

We used this approach to solve the region-coloring problem. We then discussed the A* algorithm and how it can used to find the optimal paths to the solution. We used it to build an 8-puzzle solver as well as a maze solver. In the next chapter, we will discuss genetic algorithms and how they can used to solve real-world problems.

8
Genetic Algorithms

In this chapter, we are going to learn about genetic algorithms. We will discuss the concepts of evolutionary algorithms and genetic programming, and see how they are related to genetic algorithms. We will learn about the fundamental building blocks of genetic algorithms including crossover, mutation, and fitness functions. We will then use these concepts to build various systems.

By the end of this chapter, you will know about the following:

- Understanding evolutionary and genetic algorithms
- Fundamental concepts in genetic algorithms
- Generating a bit pattern with predefined parameters
- Visualizing the progress of the evolution
- Solving the symbol regression problem
- Building an intelligent robot controller

Understanding evolutionary and genetic algorithms

A genetic algorithm is a type of evolutionary algorithm. So, in order to understand genetic algorithms, we need to discuss evolutionary algorithms. An evolutionary algorithm is a meta heuristic optimization algorithm that applies the principles of evolution to solve problems. The concept of evolution is similar to the one we find in nature. We directly use the problem's functions and variables to arrive at the solution. But in a genetic algorithm, any given problem is encoded in bit patterns that are manipulated by the algorithm.

The underlying idea in all evolutionary algorithms is that we take a population of individuals and apply the natural selection process. We start with a set of randomly selected individuals and then identify the strongest among them. The strength of each individual is determined using a **fitness function** that's predefined. In a way, we use the **survival of the fittest** approach.

We then take these selected individuals and create the next generation of individuals by recombination and mutation. We will discuss the concepts of recombination and mutation in the next section. For now, let's think of these techniques as mechanisms to create the next generation by treating the selected individuals as parents.

Once we execute recombination and mutation, we create a new set of individuals who will compete with the old ones for a place in the next generation. By discarding the weakest individuals and replacing them with offspring, we are increasing the overall fitness level of the population. We continue to iterate until the desired overall fitness is achieved.

A genetic algorithm is an evolutionary algorithm where we use a heuristic to find a string of bits that solves a problem. We continuously iterate on a population to arrive at a solution. We do this by generating new populations containing stronger individuals. We apply probabilistic operators such as **selection**, **crossover**, and **mutation** in order to generate the next generation of individuals. The individuals are basically strings, where every string is the encoded version of a potential solution.

A fitness function is used that evaluates the fitness measure of each string telling us how well suited it is to solve this problem. This fitness function is also referred to as an **evaluation function**. Genetic algorithms apply operators that are inspired from nature, which is why the nomenclature is closely related to the terms found in biology.

Fundamental concepts in genetic algorithms

In order to build a genetic algorithm, we need to understand several key concepts and terminology. These concepts are used extensively throughout the field of genetic algorithms to build solutions to various problems. One of the most important aspects of genetic algorithms is the randomness. In order to iterate, it relies on the random sampling of individuals. This means that the process is non-deterministic. So, if you run the same algorithm multiple times, you might end up with different solutions.

Let's talk about population. A population is a set of individuals that are possible candidate solutions. In a genetic algorithm, we do not maintain a single best solution at any given stage. It maintains a set of potential solutions, one of which is the best. But the other solutions play an important role during the search. Since we have a population of solutions, it is less likely that will get stuck in a local optimum. Getting stuck in the local optimum is a classic problem faced by other optimization techniques.

Now that we know about population and the stochastic nature of genetic algorithms, let's talk about the operators. In order to create the next generation of individuals, we need to make sure that they come from the strongest individuals in the current generation. Mutation is one of the ways to do it. A genetic algorithm makes random changes to one or more individuals of the current generation to yield a new candidate solution. This change is called mutation. Now this change might make that individual better or worse than existing individuals.

The next concept here is recombination, which is also called crossover. This is directly related to the role of reproduction in the evolution process. A genetic algorithm tries to combine individuals from the current generation to create a new solution. It combines some of the features of each parent individual to create this offspring. This process is called crossover. The goal is to replace the weaker individuals in the current generation with offspring generated from stronger individuals in the population.

In order to apply crossover and mutation, we need to have selection criteria. The concept of selection is inspired by the theory of natural selection. During each iteration, the genetic algorithm performs a selection process. The strongest individuals are chosen using this selection process and the weaker individuals are terminated. This is where the survival of the fittest concept comes into play. The selection process is carried out using a fitness function that computes the strength of each individual.

Generating a bit pattern with predefined parameters

Now that we know how a genetic algorithm works, let's see how to use it to solve some problems. We will be using a Python package called DEAP. You can find all the details about it at http://deap.readthedocs.io/en/master. Let's go ahead and install it by running the following command on your Terminal:

```
$ pip3 install deap
```

Genetic Algorithms

Now that the package is installed, let's quickly test it. Go into the Python shell by typing the following on your Terminal:

```
$ python3
```

Once you are inside, type the following:

```
>>> import deap
```

If you do not see an error message, we are good to go.

In this section, we will solve a variant of the **One Max problem**. The One Max problem is about generating a bit string that contains the maximum number of ones. It is a simple problem, but it's very helpful in getting familiar with the library as well as understanding how to implement solutions using genetic algorithms. In our case, we will try to generate a bit string that contains a predefined number of ones. You will see that the underlying structure and part of the code is similar to the example used in the DEAP library.

Create a new Python file and import the following:

```python
import random

from deap import base, creator, tools
```

Let's say we want to generate a bit pattern of length 75, and we want it to contain 45 ones. We need to define an evaluation function that can be used to target this objective:

```python
# Evaluation function
def eval_func(individual):
    target_sum = 45
    return len(individual) - abs(sum(individual) - target_sum),
```

If you look at the formula used in the preceding function, you can see that it reaches its maximum value when the number of ones is equal to 45. The length of each individual is 75. When the number of ones is equal to 45, the return value would be 75.

We now need to define a function to create the toolbox. Let's define a creator object for the fitness function and to keep track of the individuals. The Fitness class used here is an abstract class and it needs the weights attribute to be defined. We are building a maximizing fitness using positive weights:

```python
# Create the toolbox with the right parameters
def create_toolbox(num_bits):
    creator.create("FitnessMax", base.Fitness, weights=(1.0,))
    creator.create("Individual", list, fitness=creator.FitnessMax)
```

The first line creates a single objective maximizing fitness named `FitnessMax`. The second line deals with producing the individual. In a given process, the first individual that is created is a list of floats. In order to produce this individual, we must create an `Individual` class using the `creator`. The fitness attribute will use `FitnessMax` defined earlier.

A `toolbox` is an object that is commonly used in `DEAP` It is used to store various functions along with their arguments. Let's create this object:

```
# Initialize the toolbox
toolbox = base.Toolbox()
```

We will now start registering various functions to this `toolbox`. Let's start with the random number generator that generates a random integer between 0 and 1. This is basically to generate the bit strings:

```
# Generate attributes
toolbox.register("attr_bool", random.randint, 0, 1)
```

Let's register the `individual` function. The method `initRepeat` takes three arguments – a container class for the individual, a function used to fill the container, and the number of times we want the function to repeat itself:

```
# Initialize structures
toolbox.register("individual", tools.initRepeat, creator.Individual,
    toolbox.attr_bool, num_bits)
```

We need to register the `population` function. We want the population to be a list of individuals:

```
# Define the population to be a list of individuals
toolbox.register("population", tools.initRepeat, list,
toolbox.individual)
```

We now need to register the genetic operators. Register the evaluation function that we defined earlier, which will act as our fitness function. We want the individual, which is a bit pattern, to have 45 ones:

```
# Register the evaluation operator
toolbox.register("evaluate", eval_func)
```

Register the crossover operator named `mate` using the `cxTwoPoint` method:

```
# Register the crossover operator
toolbox.register("mate", tools.cxTwoPoint)
```

Genetic Algorithms

Register the mutation operator named mutate using `mutFlipBit`. We need to specify the probability of each attribute to be mutated using `indpb`:

```
# Register a mutation operator
toolbox.register("mutate", tools.mutFlipBit, indpb=0.05)
```

Register the selection operator using `selTournament`. It specifies which individuals will be selected for breeding:

```
# Operator for selecting individuals for breeding
toolbox.register("select", tools.selTournament, tournsize=3)
return toolbox
```

This is basically the implementation of all the concepts we discussed in the preceding section. A toolbox generator function is very common in `DEAP` and we will use it throughout this chapter. So it's important to spend some time to understand how we generated this toolbox.

Define the `main` function by starting with the length of the bit pattern:

```
if __name__ == "__main__":
    # Define the number of bits
    num_bits = 75
```

Create a toolbox using the function we defined earlier:

```
# Create a toolbox using the above parameter
toolbox = create_toolbox(num_bits)
```

We need to seed the random number generator so that we get repeatable results:

```
# Seed the random number generator
random.seed(7)
```

Create an initial population of, say, `500` individuals using the method available in the `toolbox` object. Feel free to change this number and experiment with it:

```
# Create an initial population of 500 individuals
population = toolbox.population(n=500)
```

Define the probabilities of crossing and mutating. Again, these are parameters that are defined by the user. So you can change these parameters and see how they affect the result:

```
# Define probabilities of crossing and mutating
probab_crossing, probab_mutating  = 0.5, 0.2
```

Define the number of generations that we need to iterate until the process is terminated. If you increase the number of generations, you are giving it more freedom to improve the strength of the population:

```
# Define the number of generations
num_generations = 60
```

Evaluate all the individuals in the population using the fitness functions:

```
print('\nStarting the evolution process')
# Evaluate the entire population
fitnesses = list(map(toolbox.evaluate, population))
for ind, fit in zip(population, fitnesses):
    ind.fitness.values = fit
```

Start iterating through the generations:

```
print('\nEvaluated', len(population), 'individuals')
# Iterate through generations
for g in range(num_generations):
    print("\n===== Generation", g)
```

In each generation, select the next generation individuals using the selection operator that we registered to the toolbox earlier:

```
# Select the next generation individuals
offspring = toolbox.select(population, len(population))
```

Clone the selected individuals:

```
# Clone the selected individuals
offspring = list(map(toolbox.clone, offspring))
```

Apply crossover and mutation on the next generation individuals using the probability values defined earlier. Once it's done, we need to reset the fitness values:

```
# Apply crossover and mutation on the offspring
for child1, child2 in zip(offspring[::2], offspring[1::2]):
    # Cross two individuals
    if random.random() < probab_crossing:
        toolbox.mate(child1, child2)

        # "Forget" the fitness values of the children
        del child1.fitness.values
        del child2.fitness.values
```

Apply mutation to the next generation individuals using the corresponding probability value that we defined earlier. Once it's done, reset the fitness value:

```
# Apply mutation
for mutant in offspring:
    # Mutate an individual
    if random.random() < probab_mutating:
        toolbox.mutate(mutant)
        del mutant.fitness.values
```

Evaluate the individuals with invalid fitness values:

```
# Evaluate the individuals with an invalid fitness
invalid_ind = [ind for ind in offspring if not ind.fitness.valid]
fitnesses = map(toolbox.evaluate, invalid_ind)
for ind, fit in zip(invalid_ind, fitnesses):
    ind.fitness.values = fit
print('Evaluated', len(invalid_ind), 'individuals')
```

Replace the population with the next generation individuals:

```
# The population is entirely replaced by the offspring
population[:] = offspring
```

Print the stats for the current generation to see how it's progressing:

```
# Gather all the fitnesses in one list and print the stats
fits = [ind.fitness.values[0] for ind in population]

length = len(population)
mean = sum(fits) / length
sum2 = sum(x*x for x in fits)
std = abs(sum2 / length - mean**2)**0.5

print('Min =', min(fits), ', Max =', max(fits))
print('Average =', round(mean, 2), ', Standard deviation =',
        round(std, 2))
print("\n==== End of evolution")
```

Print the final output:

```
best_ind = tools.selBest(population, 1)[0]
print('\nBest individual:\n', best_ind)
print('\nNumber of ones:', sum(best_ind))
```

The full code is given in the file `bit_counter.py`. If you run the code, you will see iterations printed to your Terminal. At the start, you will see something like the following:

```
Starting the evolution process

Evaluated 500 individuals

===== Generation 0
Evaluated 297 individuals
Min = 58.0 , Max = 75.0
Average = 70.43 , Standard deviation = 2.91

===== Generation 1
Evaluated 303 individuals
Min = 63.0 , Max = 75.0
Average = 72.44 , Standard deviation = 2.16

===== Generation 2
Evaluated 310 individuals
Min = 65.0 , Max = 75.0
Average = 73.31 , Standard deviation = 1.6

===== Generation 3
Evaluated 273 individuals
Min = 67.0 , Max = 75.0
Average = 73.76 , Standard deviation = 1.41
```

At the end, you will see something like the following that indicates the end of the evolution:

```
===== Generation 57
Evaluated 306 individuals
Min = 68.0 , Max = 75.0
Average = 74.02 , Standard deviation = 1.27

===== Generation 58
Evaluated 276 individuals
Min = 69.0 , Max = 75.0
Average = 74.15 , Standard deviation = 1.18

===== Generation 59
Evaluated 288 individuals
Min = 69.0 , Max = 75.0
Average = 74.12 , Standard deviation = 1.24

==== End of evolution

Best individual:
 [1, 1, 0, 1, 1, 0, 1, 0, 1, 0, 0, 1, 0, 1, 0, 1, 1, 1, 1, 0, 1, 0, 0, 1, 1, 1, 0, 1, 1, 1, 1, 1, 1, 1
, 1, 1, 1, 0, 0, 1, 0, 0, 1, 1, 0, 0, 1, 1, 0, 1, 1, 0, 0, 0, 1, 0, 0, 1, 1, 1, 0, 1, 1, 1, 0, 1, 1, 0, 0
, 1, 0, 0, 0, 1]

Number of ones: 45
```

Genetic Algorithms

As seen in the preceding figure, the evolution process ends after 60 generations (zero-indexed). Once it's done, the best individual is picked and printed on the output. It has 45 ones in the best individual, which is like a confirmation for us because the target sum is 45 in our evaluation function.

Visualizing the evolution

Let's see how we can visualize the evolution process. In `DEAP`, they have used a method called **Covariance Matrix Adaptation Evolution Strategy (CMA-ES)** to visualize the evolution. It is an evolutionary algorithm that's used to solve non-linear problems in the continuous domain. CMA-ES technique is robust, well studied, and is considered as state of the art in evolutionary algorithms. Let's see how it works by delving into the code provided in their source code. The following code is a slight variation of the example shown in the `DEAP` library.

Create a new Python file and import the following:

```
import numpy as np
import matplotlib.pyplot as plt
from deap import algorithms, base, benchmarks, \
        cma, creator, tools
```

Define a function to create the toolbox. We will define a `FitnessMin` function using negative weights:

```
# Function to create a toolbox
def create_toolbox(strategy):
    creator.create("FitnessMin", base.Fitness, weights=(-1.0,))
    creator.create("Individual", list, fitness=creator.FitnessMin)
```

Create the toolbox and register the evaluation function, as follows:

```
    toolbox = base.Toolbox()
    toolbox.register("evaluate", benchmarks.rastrigin)

    # Seed the random number generator
    np.random.seed(7)
```

Register the `generate` and `update` methods. This is related to the generate-update paradigm where we generate a population from a strategy and this strategy is updated based on the population:

```
toolbox.register("generate", strategy.generate, creator.Individual)
toolbox.register("update", strategy.update)

return toolbox
```

Define the `main` function. Start by defining the number of individuals and the number of generations:

```
if __name__ == "__main__":
    # Problem size
    num_individuals = 10
    num_generations = 125
```

We need to define a strategy before we start the process:

```
# Create a strategy using CMA-ES algorithm
strategy = cma.Strategy(centroid=[5.0]*num_individuals, sigma=5.0,
        lambda_=20*num_individuals)
```

Create the toolbox based on the strategy:

```
# Create toolbox based on the above strategy
toolbox = create_toolbox(strategy)
```

Create a `HallOfFame` object. The `HallOfFame` object contains the best individual that ever existed in the population. This object is kept in a sorted format at all times. This way, the first element in this object is the individual that has the best fitness value ever seen during the evolution process:

```
# Create hall of fame object
hall_of_fame = tools.HallOfFame(1)
```

Register the stats using the `Statistics` method:

```
# Register the relevant stats
stats = tools.Statistics(lambda x: x.fitness.values)
stats.register("avg", np.mean)
stats.register("std", np.std)
stats.register("min", np.min)
stats.register("max", np.max)
```

Define the `logbook` to keep track of the evolution records. It is basically a chronological list of dictionaries:

```
logbook = tools.Logbook()
logbook.header = "gen", "evals", "std", "min", "avg", "max"
```

Define objects to compile all the data:

```
# Objects that will compile the data
sigma = np.ndarray((num_generations, 1))
axis_ratio = np.ndarray((num_generations, 1))
diagD = np.ndarray((num_generations, num_individuals))
fbest = np.ndarray((num_generations, 1))
best = np.ndarray((num_generations, num_individuals))
std = np.ndarray((num_generations, num_individuals))
```

Iterate through the generations:

```
for gen in range(num_generations):
    # Generate a new population
    population = toolbox.generate()
```

Evaluate individuals using the fitness function:

```
# Evaluate the individuals
fitnesses = toolbox.map(toolbox.evaluate, population)
for ind, fit in zip(population, fitnesses):
    ind.fitness.values = fit
```

Update the strategy based on the population:

```
# Update the strategy with the evaluated individuals
toolbox.update(population)
```

Update the hall of fame and statistics with the current generation of individuals:

```
# Update the hall of fame and the statistics with the
# currently evaluated population
hall_of_fame.update(population)
record = stats.compile(population)
logbook.record(evals=len(population), gen=gen, **record)
print(logbook.stream)
```

Save the data for plotting:

```
# Save more data along the evolution for plotting
sigma[gen] = strategy.sigma
axis_ratio[gen] = max(strategy.diagD)**2/min(strategy.diagD)**2
diagD[gen, :num_individuals] = strategy.diagD**2
fbest[gen] = hall_of_fame[0].fitness.values
best[gen, :num_individuals] = hall_of_fame[0]
std[gen, :num_individuals] = np.std(population, axis=0)
```

Define the x axis and plot the stats:

```
# The x-axis will be the number of evaluations
x = list(range(0, strategy.lambda_ * num_generations,
strategy.lambda_))
avg, max_, min_ = logbook.select("avg", "max", "min")
plt.figure()
plt.semilogy(x, avg, "--b")
plt.semilogy(x, max_, "--b")
plt.semilogy(x, min_, "-b")
plt.semilogy(x, fbest, "-c")
plt.semilogy(x, sigma, "-g")
plt.semilogy(x, axis_ratio, "-r")
plt.grid(True)
plt.title("blue: f-values, green: sigma, red: axis ratio")
```

Plot the progress:

```
plt.figure()
plt.plot(x, best)
plt.grid(True)
plt.title("Object Variables")

plt.figure()
plt.semilogy(x, diagD)
plt.grid(True)
plt.title("Scaling (All Main Axes)")

plt.figure()
plt.semilogy(x, std)
plt.grid(True)
plt.title("Standard Deviations in All Coordinates")
plt.show()
```

Genetic Algorithms

The full code is given in the file `visualization.py`. If you run the code, you will see four screenshots. The first screenshot shows various parameters:

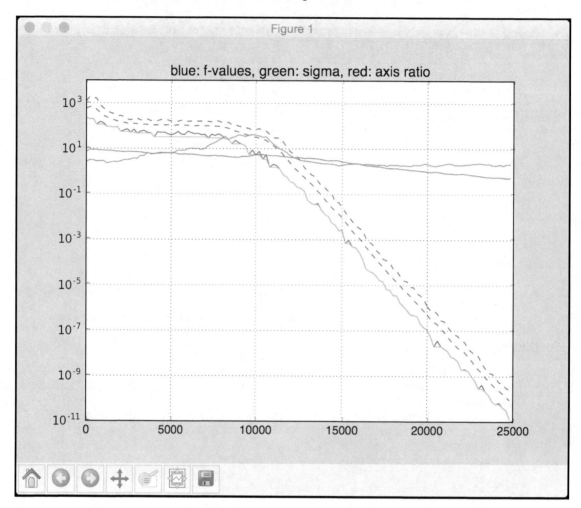

The second screenshot shows object variables:

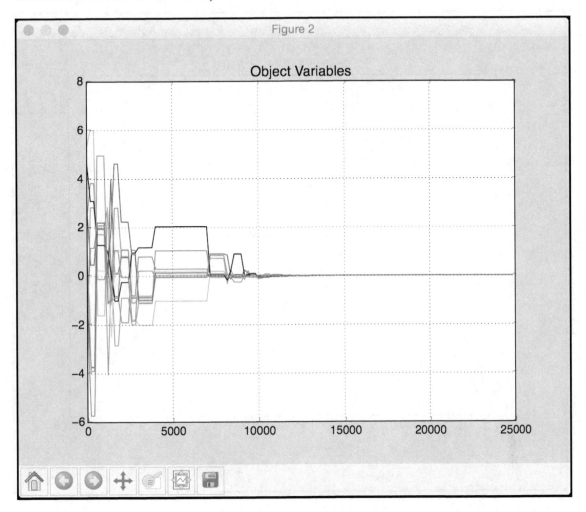

The third screenshot shows scaling:

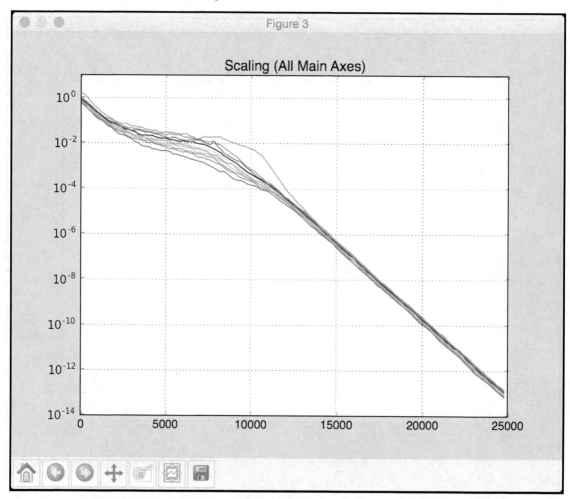

The fourth screenshot shows standard deviations:

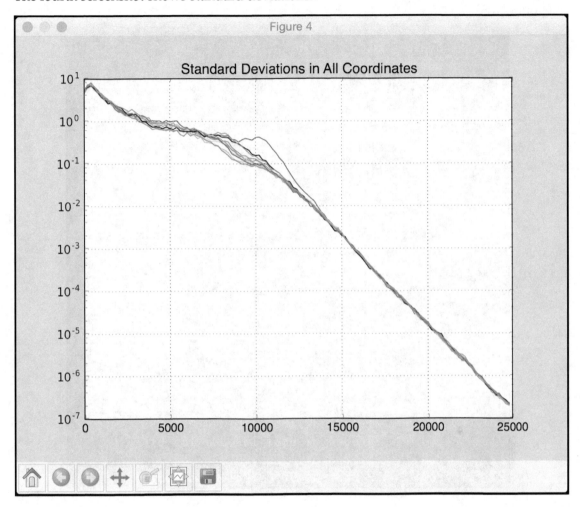

Genetic Algorithms

You will see the progress printed on the Terminal. At the start, you will see something like the following:

gen	evals	std	min	avg	max
0	200	188.36	217.082	576.281	1199.71
1	200	250.543	196.583	659.389	1869.02
2	200	273.081	199.455	683.641	1770.65
3	200	215.326	111.298	503.933	1579.3
4	200	133.046	149.47	373.124	790.899
5	200	75.4405	131.117	274.092	585.433
6	200	61.2622	91.7121	232.624	426.666
7	200	49.8303	88.8185	201.117	373.543
8	200	39.9533	85.0531	178.645	326.209
9	200	31.3781	87.4824	159.211	261.132
10	200	31.3488	54.0743	144.561	274.877
11	200	30.8796	63.6032	136.791	240.739
12	200	24.1975	70.4913	125.691	190.684
13	200	21.2274	50.6409	122.293	177.483
14	200	25.4931	67.9873	124.132	199.296
15	200	26.9804	46.3411	119.295	205.331
16	200	24.8993	56.0033	115.614	176.702
17	200	21.9789	61.4999	113.417	170.156
18	200	21.2823	50.2455	112.419	190.677
19	200	22.5016	48.153	111.543	166.2
20	200	21.1602	32.1864	106.044	171.899
21	200	23.3864	52.8601	107.301	163.617
22	200	23.1008	51.1226	109.628	185.777
23	200	22.0836	51.3058	106.402	179.673

At the end, you will see the following:

```
100    200    2.38865e-07    1.12678e-07    5.18814e-07    1.23527e-06
101    200    1.49444e-07    5.56979e-08    3.3199e-07     7.98774e-07
102    200    1.11635e-07    2.07109e-08    2.41361e-07    7.96738e-07
103    200    9.50257e-08    3.69117e-08    1.94641e-07    5.75896e-07
104    200    5.63849e-08    2.09827e-08    1.26148e-07    2.887e-07
105    200    4.42488e-08    1.64212e-08    8.6972e-08     2.58639e-07
106    200    2.34933e-08    1.28302e-08    5.47789e-08    1.54658e-07
107    200    1.74434e-08    7.13185e-09    3.64705e-08    9.88235e-08
108    200    1.17157e-08    6.32208e-09    2.54673e-08    7.13075e-08
109    200    8.73027e-09    4.60369e-09    1.79681e-08    5.88066e-08
110    200    6.39874e-09    1.92573e-09    1.43229e-08    4.00087e-08
111    200    5.31196e-09    2.05551e-09    1.13736e-08    3.16793e-08
112    200    3.15607e-09    1.72427e-09    7.28548e-09    1.67727e-08
113    200    2.3789e-09     1.01164e-09    5.01177e-09    1.24541e-08
114    200    1.38424e-09    6.43112e-10    2.94696e-09    9.25819e-09
115    200    1.04172e-09    2.87571e-10    2.06068e-09    7.90436e-09
116    200    6.08685e-10    4.32905e-10    1.4704e-09     3.80221e-09
117    200    4.51515e-10    2.1538e-10     9.23627e-10    2.2759e-09
118    200    2.77204e-10    1.46869e-10    6.3507e-10     1.44637e-09
119    200    2.06475e-10    7.54881e-11    4.41427e-10    1.33167e-09
120    200    1.3138e-10     5.97282e-11    2.98116e-10    8.60453e-10
121    200    9.52385e-11    6.753e-11      2.32358e-10    5.45441e-10
122    200    7.55001e-11    4.1851e-11     1.72688e-10    5.05054e-10
123    200    5.52125e-11    3.2216e-11     1.23505e-10    3.10081e-10
124    200    4.38068e-11    1.32871e-11    8.94929e-11    2.57202e-10
```

As seen from the preceding figure, the values keep decreasing as we progress. This indicates that it's converging.

Solving the symbol regression problem

Let's see how to use genetic programming to solve the symbol regression problem. It is important to understand that genetic programming is not the same as genetic algorithms. Genetic programming is a type of evolutionary algorithm in which the solutions occur in the form of computer programs. Basically, the individuals in each generation would be computer programs and their fitness level corresponds to their ability to solve problems. These programs are modified, at each iteration, using a genetic algorithm. In essence, genetic programming is the application of a genetic algorithm.

Genetic Algorithms

Coming to the symbol regression problem, we have a polynomial expression that needs to be approximated here. It's a classic regression problem where we try to estimate the underlying function. In this example, we will use the expression: $f(x) = 2x^3 - 3x^2 + 4x - 1$

The code discussed here is a variant of the symbol regression problem given in the DEAP library. Create a new Python file and import the following:

```python
import operator
import math
import random

import numpy as np
from deap import algorithms, base, creator, tools, gp
```

Create a division operator that can handle divide-by-zero error gracefully:

```python
# Define new functions
def division_operator(numerator, denominator):
    if denominator == 0:
        return 1

    return numerator / denominator
```

Define the evaluation function that will be used for fitness calculation. We need to define a callable function to run computations on the input individual:

```python
# Define the evaluation function
def eval_func(individual, points):
    # Transform the tree expression in a callable function
    func = toolbox.compile(expr=individual)
```

Compute the **mean squared error** (**MSE**) between the function defined earlier and the original expression:

```python
    # Evaluate the mean squared error
    mse = ((func(x) - (2 * x**3 - 3 * x**2 - 4 * x + 1))**2 for x in points)

    return math.fsum(mse) / len(points),
```

Define a function to create the toolbox. In order to create the toolbox here, we need to create a set of primitives. These primitives are basically operators that will be used during the evolution. They serve as building blocks for the individuals. We are going to use basic arithmetic functions as our primitives here:

```
# Function to create the toolbox
def create_toolbox():
    pset = gp.PrimitiveSet("MAIN", 1)
    pset.addPrimitive(operator.add, 2)
    pset.addPrimitive(operator.sub, 2)
    pset.addPrimitive(operator.mul, 2)
    pset.addPrimitive(division_operator, 2)
    pset.addPrimitive(operator.neg, 1)
    pset.addPrimitive(math.cos, 1)
    pset.addPrimitive(math.sin, 1)
```

We now need to declare an ephemeral constant. It is a special terminal type that does not have a fixed value. When a given program appends such an ephemeral constant to the tree, the function gets executed. The result is then inserted into the tree as a constant terminal. These constant terminals can take the values -1, 0 or 1:

```
    pset.addEphemeralConstant("rand101", lambda: random.randint(-1,1))
```

The default name for the arguments is ARGx. Let's rename it x. It's not exactly necessary, but it's a useful feature that comes in handy:

```
    pset.renameArguments(ARG0='x')
```

We need to define two object types – fitness and an individual. Let's do it using the creator:

```
    creator.create("FitnessMin", base.Fitness, weights=(-1.0,))
    creator.create("Individual", gp.PrimitiveTree, 
fitness=creator.FitnessMin)
```

Create the `toolbox` and `register` the functions. The registration process is similar to previous sections:

```
    toolbox = base.Toolbox()

    toolbox.register("expr", gp.genHalfAndHalf, pset=pset, min_=1, max_=2)
    toolbox.register("individual", tools.initIterate, creator.Individual, 
toolbox.expr)
    toolbox.register("population", tools.initRepeat, list, 
toolbox.individual)
    toolbox.register("compile", gp.compile, pset=pset)
    toolbox.register("evaluate", eval_func, points=[x/10. for x in
```

Genetic Algorithms

```
range(-10,10)])
    toolbox.register("select", tools.selTournament, tournsize=3)
    toolbox.register("mate", gp.cxOnePoint)
    toolbox.register("expr_mut", gp.genFull, min_=0, max_=2)
    toolbox.register("mutate", gp.mutUniform, expr=toolbox.expr_mut,
pset=pset)

    toolbox.decorate("mate",
gp.staticLimit(key=operator.attrgetter("height"), max_value=17))
    toolbox.decorate("mutate",
gp.staticLimit(key=operator.attrgetter("height"), max_value=17))

    return toolbox
```

Define the `main` function and start by seeding the random number generator:

```
if __name__ == "__main__":
    random.seed(7)
```

Create the `toolbox` object:

```
toolbox = create_toolbox()
```

Define the initial population using the method available in the `toolbox` object. We will use 450 individuals. The user defines this number, so you should feel free to experiment with it. Also define the `hall_of_fame` object:

```
population = toolbox.population(n=450)
hall_of_fame = tools.HallOfFame(1)
```

Statistics are useful when we build genetic algorithms. Define the stats objects:

```
stats_fit = tools.Statistics(lambda x: x.fitness.values)
stats_size = tools.Statistics(len)
```

Register the stats using the objects defined previously:

```
mstats = tools.MultiStatistics(fitness=stats_fit, size=stats_size)
mstats.register("avg", np.mean)
mstats.register("std", np.std)
mstats.register("min", np.min)
mstats.register("max", np.max)
```

Define the crossover probability, mutation probability, and the number of generations:

```
probab_crossover = 0.4
probab_mutate = 0.2
num_generations = 60
```

Run the evolutionary algorithm using the above parameters:

```
population, log = algorithms.eaSimple(population, toolbox,
        probab_crossover, probab_mutate, num_generations,
        stats=mstats, halloffame=hall_of_fame, verbose=True)
```

The full code is given in the file `symbol_regression.py`. If you run the code, you will see the following on your Terminal at the start of the evolution:

		fitness				size			
gen	nevals	avg	max	min	std	avg	max	min	std
0	450	18.6918	47.1923	7.39087	6.27543	3.73556	7	2	1.62449
1	251	15.4572	41.3823	4.46965	4.54993	3.80222	12	1	1.81316
2	236	13.2545	37.7223	4.46965	4.06145	3.96889	12	1	1.98861
3	251	12.2299	60.828	4.46965	4.70055	4.19556	12	1	1.9971
4	235	11.001	47.1923	4.46965	4.48841	4.84222	13	1	2.17245
5	229	9.44483	31.478	4.46965	3.8796	5.56	19	1	2.43168
6	225	8.35975	22.0546	3.02133	3.40547	6.38889	15	1	2.40875
7	237	7.99309	31.1356	1.81133	4.08463	7.14667	16	1	2.57782
8	224	7.42611	359.418	1.17558	17.0167	8.33333	19	1	3.11127
9	237	5.70308	24.1921	1.17558	3.71991	9.64444	23	1	3.31365
10	254	5.27991	30.4315	1.13301	4.13556	10.5089	25	1	3.51898

At the end, you will see the following:

36	209	1.10464	22.0546	0.0474957	2.71898	26.4867	46	1	5.23289
37	258	1.61958	86.0936	0.0382386	6.1839	27.2111	45	3	4.75557
38	257	2.03651	70.4768	0.0342642	5.15243	26.5311	49	1	6.22327
39	235	1.95531	185.328	0.0472693	9.32516	26.9711	48	1	6.00345
40	234	1.51403	28.5529	0.0472693	3.24513	26.6867	52	1	5.39811
41	230	1.4753	70.4768	0.0472693	5.4607	27.1	46	3	4.7433
42	233	12.3648	4880.09	0.0396503	229.754	26.88	53	1	5.18192
43	251	1.807	86.0936	0.0396503	5.85281	26.4889	50	1	5.43741
44	236	9.30096	3481.25	0.0277886	163.888	26.9622	55	1	6.27169
45	231	1.73196	86.7372	0.0342642	6.8119	27.4711	51	2	5.27807
46	227	1.86086	185.328	0.0342642	10.1143	28.0644	56	1	6.10812
47	216	12.5214	4923.66	0.0342642	231.837	29.1022	54	1	6.45898
48	232	14.3469	5830.89	0.0322462	274.536	29.8244	58	3	6.24093
49	242	2.56984	272.833	0.0322462	18.2752	29.9267	51	1	6.31446
50	227	2.80136	356.613	0.0322462	21.0416	29.7978	56	4	6.50275
51	243	1.75099	86.0936	0.0322462	5.70833	29.8089	56	1	6.62379
52	253	10.9184	3435.84	0.0227048	163.602	29.9911	55	1	6.66833
53	243	1.80265	48.0418	0.0227048	4.73856	29.88	55	1	7.33084
54	234	1.74487	86.0936	0.0227048	6.0249	30.6067	55	1	6.85782
55	220	1.58888	31.094	0.0132398	3.82809	30.5644	54	1	6.96669
56	234	1.46711	103.287	0.00766444	6.81157	30.6689	55	3	6.6806
57	250	17.0896	6544.17	0.00424267	308.689	31.1267	60	4	7.25837
58	231	1.66757	141.584	0.00144401	7.35306	32	52	1	7.23295
59	229	2.22325	265.224	0.00144401	13.388	33.5489	64	1	8.38351
60	248	2.60303	521.804	0.00144401	24.7018	35.2533	58	1	7.61506

Building an intelligent robot controller

Let's see how to build a robot controller using a genetic algorithm. We are given a map with the targets sprinkled all over it. The map looks like this:

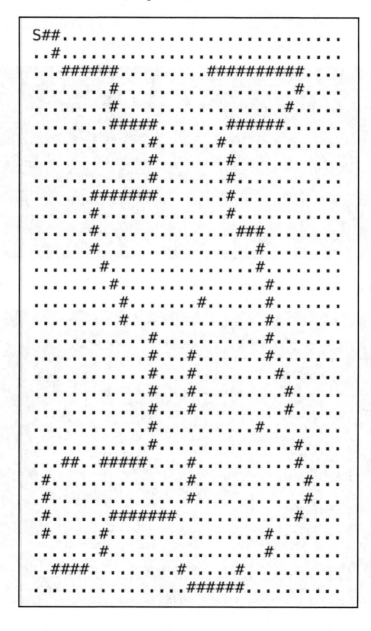

There are 124 targets in the preceding map. The goal of the robot controller is to automatically traverse the map and consume all those targets. This program is a variant of the artificial ant program given in the `deap` library.

Create a new Python file and import the following:

```
import copy
import random
from functools import partial

import numpy as np
from deap import algorithms, base, creator, tools, gp
```

Create the class to control the robot:

```
class RobotController(object):
    def __init__(self, max_moves):
        self.max_moves = max_moves
        self.moves = 0
        self.consumed = 0
        self.routine = None
```

Define the directions and movements:

```
        self.direction = ["north", "east", "south", "west"]
        self.direction_row = [1, 0, -1, 0]
        self.direction_col = [0, 1, 0, -1]
```

Define the reset functionality:

```
    def _reset(self):
        self.row = self.row_start
        self.col = self.col_start
        self.direction = 1
        self.moves = 0
        self.consumed = 0
        self.matrix_exc = copy.deepcopy(self.matrix)
```

Define the conditional operator:

```
    def _conditional(self, condition, out1, out2):
        out1() if condition() else out2()
```

[225]

Genetic Algorithms

Define the left turning operator:

```python
def turn_left(self):
    if self.moves < self.max_moves:
        self.moves += 1
        self.direction = (self.direction - 1) % 4
```

Define the right turning operator:

```python
def turn_right(self):
    if self.moves < self.max_moves:
        self.moves += 1
        self.direction = (self.direction + 1) % 4
```

Define the method to control how the robot moves forward:

```python
def move_forward(self):
    if self.moves < self.max_moves:
        self.moves += 1
        self.row = (self.row + self.direction_row[self.direction]) % 
            self.matrix_row
        self.col = (self.col + self.direction_col[self.direction]) % 
            self.matrix_col

        if self.matrix_exc[self.row][self.col] == "target":
            self.consumed += 1

        self.matrix_exc[self.row][self.col] = "passed"
```

Define a method to sense the target. If you see the target ahead, then update the matrix accordingly:

```python
def sense_target(self):
    ahead_row = (self.row + self.direction_row[self.direction]) % 
  self.matrix_row
    ahead_col = (self.col + self.direction_col[self.direction]) % 
  self.matrix_col
        return self.matrix_exc[ahead_row][ahead_col] == "target"
```

If you see the target ahead, then create the relevant function and return it:

```python
def if_target_ahead(self, out1, out2):
    return partial(self._conditional, self.sense_target, out1, out2)
```

Define the method to run it:

```
def run(self, routine):
    self._reset()
    while self.moves < self.max_moves:
        routine()
```

Define a function to traverse the input map. The symbol # indicates all the targets on the map and the symbol S indicates the starting point. The symbol . denotes empty cells:

```
def traverse_map(self, matrix):
    self.matrix = list()
    for i, line in enumerate(matrix):
        self.matrix.append(list())

        for j, col in enumerate(line):
            if col == "#":
                self.matrix[-1].append("target")

            elif col == ".":
                self.matrix[-1].append("empty")

            elif col == "S":
                self.matrix[-1].append("empty")
                self.row_start = self.row = i
                self.col_start = self.col = j
                self.direction = 1

    self.matrix_row = len(self.matrix)
    self.matrix_col = len(self.matrix[0])
    self.matrix_exc = copy.deepcopy(self.matrix)
```

Define a class to generate functions depending on the number of input arguments:

```
class Prog(object):
    def _progn(self, *args):
        for arg in args:
            arg()

    def prog2(self, out1, out2):
        return partial(self._progn, out1, out2)

    def prog3(self, out1, out2, out3):
        return partial(self._progn, out1, out2, out3)
```

Genetic Algorithms

Define an evaluation function for each individual:

```
def eval_func(individual):
    global robot, pset

    # Transform the tree expression to functional Python code
    routine = gp.compile(individual, pset)
```

Run the current program:

```
    # Run the generated routine
    robot.run(routine)
    return robot.consumed,
```

Define a function to create the toolbox and add primitives:

```
def create_toolbox():
    global robot, pset

    pset = gp.PrimitiveSet("MAIN", 0)
    pset.addPrimitive(robot.if_target_ahead, 2)
    pset.addPrimitive(Prog().prog2, 2)
    pset.addPrimitive(Prog().prog3, 3)
    pset.addTerminal(robot.move_forward)
    pset.addTerminal(robot.turn_left)
    pset.addTerminal(robot.turn_right)
```

Create the object types using the fitness function:

```
    creator.create("FitnessMax", base.Fitness, weights=(1.0,))
    creator.create("Individual", gp.PrimitiveTree,
fitness=creator.FitnessMax)
```

Create the `toolbox` and `register` all the operators:

```
    toolbox = base.Toolbox()

    # Attribute generator
    toolbox.register("expr_init", gp.genFull, pset=pset, min_=1, max_=2)

    # Structure initializers
    toolbox.register("individual", tools.initIterate, creator.Individual,
toolbox.expr_init)
    toolbox.register("population", tools.initRepeat, list,
toolbox.individual)

    toolbox.register("evaluate", eval_func)
    toolbox.register("select", tools.selTournament, tournsize=7)
    toolbox.register("mate", gp.cxOnePoint)
```

```
    toolbox.register("expr_mut", gp.genFull, min_=0, max_=2)
    toolbox.register("mutate", gp.mutUniform, expr=toolbox.expr_mut,
pset=pset)

    return toolbox
```

Define the `main` function and start by seeding the random number generator:

```
if __name__ == "__main__":
    global robot

    # Seed the random number generator
    random.seed(7)
```

Create the robot controller object using the initialization parameter:

```
    # Define the maximum number of moves
    max_moves = 750

    # Create the robot object
    robot = RobotController(max_moves)
```

Create the `toolbox` using the function we defined earlier:

```
    # Create the toolbox
    toolbox = create_toolbox()
```

Read the map data from the input file:

```
    # Read the map data
    with open('target_map.txt', 'r') as f:
        robot.traverse_map(f)
```

Define the population with 400 individuals and define the `hall_of_fame` object:

```
    # Define population and hall of fame variables
    population = toolbox.population(n=400)
    hall_of_fame = tools.HallOfFame(1)
```

Register the stats:

```
    # Register the stats
    stats = tools.Statistics(lambda x: x.fitness.values)
    stats.register("avg", np.mean)
    stats.register("std", np.std)
    stats.register("min", np.min)
    stats.register("max", np.max)
```

Define the crossover probability, mutation probability, and the number of generations:

```
# Define parameters
probab_crossover = 0.4
probab_mutate = 0.3
num_generations = 50
```

Run the evolutionary algorithm using the parameters defined earlier:

```
# Run the algorithm to solve the problem
algorithms.eaSimple(population, toolbox, probab_crossover,
    probab_mutate, num_generations, stats,
    halloffame=hall_of_fame)
```

The full code is given in the file `robot.py`. If you run the code, you will get the following on your Terminal:

```
gen    nevals  avg      std      min  max
0      400     1.4875   4.37491  0    62
1      231     4.285    7.56993  0    73
2      235     10.8925  14.8493  0    73
3      231     21.72    22.1239  0    73
4      238     29.9775  27.7861  0    76
5      224     37.6275  31.8698  0    76
6      231     42.845   33.0541  0    80
7      223     43.55    33.9369  0    83
8      234     44.0675  34.5201  0    83
9      231     49.2975  34.3065  0    83
10     249     47.075   36.4106  0    93
11     222     52.7925  36.2826  0    97
12     248     51.0725  37.2598  0    97
13     234     54.01    37.4614  0    97
14     229     59.615   37.7894  0    97
15     228     63.3     39.8205  0    97
16     220     64.605   40.3962  0    97
17     236     62.545   40.5607  0    97
18     233     67.99    38.9033  0    97
19     236     66.4025  39.6574  0    97
20     221     69.785   38.7117  0    97
21     244     65.705   39.0957  0    97
22     230     70.32    37.1206  0    97
23     241     67.3825  39.4028  0    97
```

Towards the end, you will see the following:

```
26    214    71.505   36.964   0    97
27    246    72.72    37.1637  0    97
28    238    73.5975  36.5385  0    97
29    239    76.405   35.5696  0    97
30    246    78.6025  33.4281  0    97
31    240    74.83    36.5157  0    97
32    216    80.2625  32.6659  0    97
33    220    80.6425  33.0933  0    97
34    247    78.245   34.6022  0    97
35    241    81.22    32.1885  0    97
36    234    83.6375  29.0002  0    97
37    228    82.485   31.7354  0    97
38    219    83.4625  30.0592  0    97
39    212    88.64    24.2702  0    97
40    231    86.7275  27.0879  0    97
41    229    89.1825  23.8773  0    97
42    216    87.96    25.1649  0    97
43    218    86.85    27.1116  0    97
44    236    88.78    23.7278  0    97
45    225    89.115   23.4212  0    97
46    232    88.5425  24.187   0    97
47    245    87.7775  25.3909  0    97
48    231    87.78    26.3786  0    97
49    238    88.8525  24.5115  0    97
50    233    87.82    25.4164  1    97
```

Summary

In this chapter, we learned about genetic algorithms and their underlying concepts. We discussed evolutionary algorithms and genetic programming. We understood how they are related to genetic algorithms. We discussed the fundamental building blocks of genetic algorithms including the concepts of population, crossover, mutation, selection, and fitness function. We learned how to generate a bit pattern with predefined parameters. We discussed how to visualize the evolution process using CMA-ES. We learnt how to solve the symbol regression problem in this paradigm. We then used these concepts to build a robot controller to traverse a map and consume all the targets. In the next chapter, we will learn about reinforcement learning and see how to build a smart agent.

9
Building Games With Artificial Intelligence

In this chapter, we are going to learn how to build games with Artificial Intelligence. We will learn how to use search algorithms to effectively come up with strategies to win the games. We will then use these algorithms to build intelligent bots for different games.

By the end of this chapter, you will understand the following concepts:

- Using search algorithms in games
- Combinatorial search
- Minimax algorithm
- Alpha-Beta pruning
- Negamax algorithm
- Building a bot to play Last Coin Standing
- Building a bot to play Tic Tac Toe
- Building two bots to play Connect Four against each other
- Building two bots to play Hexapawn against each other

Using search algorithms in games

Search algorithms are used in games to figure out a strategy. The algorithms search through the possibilities and pick the best move. There are various parameters to think about – speed, accuracy, complexity, and so on. These algorithms consider all possible actions available at this time and then evaluate their future moves based on these options. The goal of these algorithms is to find the optimal set of moves that will help them arrive at the final condition. Every game has a different set of winning conditions. These algorithms use those conditions to find the set of moves.

The description given in the previous paragraph is ideal if there is no opposing player. Things are not as straightforward with games that have multiple players. Let's consider a two-player game. For every move made by a player, the opposing player will make a move to prevent the player from achieving the goal. So when a search algorithm finds the optimal set of moves from the current state, it cannot just go ahead and make those moves because the opposing player will stop it. This basically means that search algorithms need to constantly re-evaluate after each move.

Let's discuss how a computer perceives any given game. We can think of a game as a search tree. Each node in this tree represents a future state. For example, if you are playing **Tic–Tac–Toe** (Noughts and Crosses), you can construct this tree to represent all possible moves. We start from the root of the tree, which is the starting point of the game. This node will have several children that represent various possible moves. Those children, in turn, will have more children that represent game states after more moves by the opponent. The terminal nodes of the tree represent the final results of the game after making various moves. The game would either end in a draw or one of the players would win it. The search algorithms search through this tree to make decisions at each step of the game.

Combinatorial search

Search algorithms appear to solve the problem of adding intelligence to games, but there's a drawback. These algorithms employ a type of search called exhaustive search, which is also known as brute force search. It basically explores the entire search space and tests every possible solution. It means that, in the worst case, we will have to explore all the possible solutions before we get the right solution.

As the games get more complex, we cannot rely on brute force search because the number of possibilities gets enormous. This quickly becomes computationally intractable. In order to solve this problem, we use combinatorial search to solve problems. It refers to a field of study where search algorithms efficiently explore the solution space using heuristics or by reducing the size of the search space. This is very useful in games like Chess or Go. Combinatorial search works efficiently by using pruning strategies. These strategies help it avoid testing all possible solutions by eliminating the ones that are obviously wrong. This helps save time and effort.

Minimax algorithm

Now that we have briefly discussed combinatorial search, let's talk about the heuristics that are employed by combinatorial search algorithms. These heuristics are used to speed up the search strategy and the Minimax algorithm is one such strategy used by combinatorial search. When two players are playing against each other, they are basically working towards opposite goals. So each side needs to predict what the opposing player is going to do in order to win the game. Keeping this in mind, Minimax tries to achieve this through strategy. It will try to minimize the function that the opponent is trying to maximize.

As we know, brute forcing the solution is not an option. The computer cannot go through all the possible states and then get the best possible set of moves to win the game. The computer can only optimize the moves based on the current state using a heuristic. The computer constructs a tree and it starts from the bottom. It evaluates which moves would benefit its opponent. Basically, it knows which moves the opponent is going to make based on the premise that the opponent will make the moves that would benefit them the most, and thereby be of the least benefit to the computer. This outcome is one of the terminal nodes of the tree and the computer uses this position to work backwards. Each option that's available to the computer can be assigned a value and it can then pick the highest value to take an action.

Alpha-Beta pruning

Minimax search is an efficient strategy, but it still ends up exploring parts of the tree that are irrelevant. Let's consider a tree where we are supposed to search for solutions. Once we find an indicator on a node that tells us that the solution does not exist in that sub-tree, there is no need to evaluate that sub-tree. But Minimax search is a bit too conservative, so it ends up exploring that sub-tree.

We need to be smart about it and avoid searching a part of a tree that is not necessary. This process is called **pruning** and Alpha-Beta pruning is a type of avoidance strategy that is used to avoid searching parts of the tree that do not contain the solution.

The Alpha and Beta parameters in alpha-beta pruning refer to the two bounds that are used during the calculation. These parameters refer to the values that restrict the set of possible solutions. This is based on the section of the tree that has already been explored. Alpha is the maximum lower bound of the number of possible solutions and Beta is the minimum upper bound on the number of possible solutions.

As we discussed earlier, each node can be assigned a value based on the current state. When the algorithm considers any new node as a potential path to the solution, it can work out if the current estimate of the value of the node lies between alpha and beta. This is how it prunes the search.

Negamax algorithm

The **Negamax** algorithm is a variant of Minimax that's frequently used in real world implementations. A two-player game is usually a zero-sum game, which means that one player's loss is equal to another player's gain and vice versa. Negamax uses this property extensively to come up with a strategy to increases its chances of winning the game.

In terms of the game, the value of a given position to the first player is the negation of the value to the second player. Each player looks for a move that will maximize the damage to the opponent. The value resulting from the move should be such that the opponent gets the least value. This works both ways seamlessly, which means that a single method can be used to value the positions. This is where it has an advantage over Minimax in terms of simplicity. Minimax requires that the first player select the move with the maximum value, whereas the second player must select a move with the minimum value. Alpha-beta pruning is used here as well.

Installing easyAI library

We will be using a library called `easyAI` in this chapter. It is an artificial intelligence framework and it provides all the functionality necessary to build two-player games. You can learn about it at http://zulko.github.io/easyAI.

Install it by running the following command on your Terminal:

```
$ pip3 install easyAI
```

We need some of the files to be accessible in order to use some of the pre-built routines. For ease of use, the code provided with this book contains a folder called easyAI. Make sure you place this folder in the same folder as your code files. This folder is basically a subset of the easyAI GitHub repository available at https://github.com/Zulko/easyAI. You can go through the source code to make yourself more familiar with it.

Building a bot to play Last Coin Standing

This is a game where we have a pile of coins and each player takes turns to take a number of coins from the pile. There is a lower and an upper bound on the number of coins that can be taken from the pile. The goal of the game is to avoid taking the last coin in the pile. This recipe is a variant of the Game of Bones recipe given in the easyAI library. Let's see how to build a game where the computer can play against the user.

Create a new Python file and import the following packages:

```
from easyAI import TwoPlayersGame, id_solve, Human_Player, AI_Player
from easyAI.AI import TT
```

Create a class to handle all the operations of the game. We will be inheriting from the base class TwoPlayersGame available in the easyAI library. There are a couple of parameters that have been to defined in order for it to function properly. The first one is the players variable. We will talk about the player object later. Create the class using the following code:

```
class LastCoinStanding(TwoPlayersGame):
    def __init__(self, players):
        # Define the players. Necessary parameter.
        self.players = players
```

Define who is going to start the game. The players are numbered from one. So in this case, player one starts the game:

```
        # Define who starts the game. Necessary parameter.
        self.nplayer = 1
```

Define the number of coins in the pile. You are free to choose any number here. In this case, let's choose 25:

```
# Overall number of coins in the pile
self.num_coins = 25
```

Define the maximum number of coins that can be taken out in any move. You are free to choose any number for this parameter as well. Let's choose 4 in our case:

```
# Define max number of coins per move
self.max_coins = 4
```

Define all the possible movies. In this case, players can take either 1, 2, 3, or 4 coins in each move:

```
# Define possible moves
def possible_moves(self):
    return [str(x) for x in range(1, self.max_coins + 1)]
```

Define a method to remove the coins and keep track of the number of coins remaining in the pile:

```
# Remove coins
def make_move(self, move):
    self.num_coins -= int(move)
```

Check if somebody won the game by checking the number of coins remaining:

```
# Did the opponent take the last coin?
def win(self):
    return self.num_coins <= 0
```

Stop the game after somebody wins it:

```
# Stop the game when somebody wins
def is_over(self):
    return self.win()
```

Compute the score based on the `win` method. It's necessary to define this method:

```
# Compute score
def scoring(self):
    return 100 if self.win() else 0
```

Define a method to show the current status of the pile:

```
# Show number of coins remaining in the pile
def show(self):
    print(self.num_coins, 'coins left in the pile')
```

Define the `main` function and start by defining the transposition table. Transposition tables are used in games to store the positions and movements so as to speed up the algorithm. Type in the following code:

```
if __name__ == "__main__":
    # Define the transposition table
    tt = TT()
```

Define the method `ttentry` to get the number of coins. It's an optional method that's used to create a string to describe the game:

```
# Define the method
LastCoinStanding.ttentry = lambda self: self.num_coins
```

Let's solve the game using AI. The function `id_solve` is used to solve a given game using iterative deepening. It basically determines who can win a game using all the paths. It looks to answer questions such as, Can the first player force a win by playing perfectly? Will the computer always lose against a perfect opponent?

The method `id_solve` explores various options in the game's Negamax algorithm several times. It always starts at the initial state of the game and takes increasing depth to keep going. It will do it until the score indicates that somebody will win or lose. The second argument in the method takes a list of depths that it will try out. In this case, it will try all the values from 2 to 20:

```
# Solve the game
result, depth, move = id_solve(LastCoinStanding,
        range(2, 20), win_score=100, tt=tt)
print(result, depth, move)
```

Start the game against the computer:

```
# Start the game
game = LastCoinStanding([AI_Player(tt), Human_Player()])
game.play()
```

The full code is given in the file `coins.py`. It's an interactive program, so it will expect input from the user. If you run the code, you will basically be playing against the computer. Your goal is to force the computer to take the last coin, so that you win the game. If you run the code, you will get the following output on your Terminal at the beginning:

```
d:2, a:0, m:1
d:3, a:0, m:1
d:4, a:0, m:1
d:5, a:0, m:1
d:6, a:0, m:1
d:7, a:0, m:1
d:8, a:0, m:1
d:9, a:0, m:1
d:10, a:100, m:4
1 10 4
25 coins left in the pile

Move #1: player 1 plays 4 :
21 coins left in the pile

Player 2 what do you play ? 1

Move #2: player 2 plays 1 :
20 coins left in the pile

Move #3: player 1 plays 4 :
16 coins left in the pile
```

If you scroll down, you will see the following towards the end:

```
Move #5: player 1 plays 2 :
11 coins left in the pile

Player 2 what do you play ? 4

Move #6: player 2 plays 4 :
7 coins left in the pile

Move #7: player 1 plays 1 :
6 coins left in the pile

Player 2 what do you play ? 2

Move #8: player 2 plays 2 :
4 coins left in the pile

Move #9: player 1 plays 3 :
1 coins left in the pile

Player 2 what do you play ? 1

Move #10: player 2 plays 1 :
0 coins left in the pile
```

As we can see, the computer wins the game because the user picked up the last coin.

Building a bot to play Tic-Tac-Toe

Tic-Tac-Toe (Noughts and Crosses) is probably one of the most famous games. Let's see how to build a game where the computer can play against the user. This is a minor variant of the Tic-Tac-Toe recipe given in the `easyAI` library.

Create a new Python file and import the following packages:

```
from easyAI import TwoPlayersGame, AI_Player, Negamax
from easyAI.Player import Human_Player
```

Define a class that contains all the methods to play the game. Start by defining the players and who starts the game:

```
class GameController(TwoPlayersGame):
    def __init__(self, players):
        # Define the players
        self.players = players

        # Define who starts the game
        self.nplayer = 1
```

We will be using a 3×3 board numbered from one to nine row-wise:

```
        # Define the board
        self.board = [0] * 9
```

Define a method to compute all the possible moves:

```
    # Define possible moves
    def possible_moves(self):
        return [a + 1 for a, b in enumerate(self.board) if b == 0]
```

Define a method to update the board after making a move:

```
    # Make a move
    def make_move(self, move):
        self.board[int(move) - 1] = self.nplayer
```

Define a method to see if somebody has lost the game. We will be checking if somebody has three in a row:

```python
# Does the opponent have three in a line?
def loss_condition(self):
    possible_combinations = [[1,2,3], [4,5,6], [7,8,9],
            [1,4,7], [2,5,8], [3,6,9], [1,5,9], [3,5,7]]
    return any([all([(self.board[i-1] == self.nopponent)
            for i in combination]) for combination in possible_combinations])
```

Check if the game is over using the `loss_condition` method:

```python
# Check if the game is over
def is_over(self):
    return (self.possible_moves() == []) or self.loss_condition()
```

Define a method to show the current progress:

```python
# Show current position
def show(self):
    print('\n'+'\n'.join([' '.join([['. ', 'O', 'X'][self.board[3*j + i]]
            for i in range(3)]) for j in range(3)]))
```

Compute the score using the `loss_condition` method:

```python
# Compute the score
def scoring(self):
    return -100 if self.loss_condition() else 0
```

Define the main function and start by defining the algorithm. We will be using Negamax as the AI algorithm for this game. We can specify the number of steps in advance that the algorithm should think. In this case, let's choose 7:

```python
if __name__ == "__main__":
    # Define the algorithm
    algorithm = Negamax(7)
```

Start the game:

```python
# Start the game
GameController([Human_Player(), AI_Player(algorithm)]).play()
```

The full code is given in the file `tic_tac_toe.py`. It's an interactive game where you play against the computer. If you run the code, you will get the following output on your Terminal at the beginning:

```
. . .
. . .
. . .
Player 1 what do you play ? 5
Move #1: player 1 plays 5 :
. . .
. O .
. . .
Move #2: player 2 plays 1 :
X . .
. O .
. . .
Player 1 what do you play ? 9
Move #3: player 1 plays 9 :
X . .
. O .
. . O
```

If you scroll down, you will see the following printed on your Terminal once it finishes executing the code:

```
X O X
. O .
. X O
Player 1 what do you play ? 4
Move #7: player 1 plays 4 :
X O X
O O .
. X O
Move #8: player 2 plays 6 :
X O X
O O X
. X O
Player 1 what do you play ? 7
Move #9: player 1 plays 7 :
X O X
O O X
O X O
```

As we can see, the game ends in a draw.

Building two bots to play Connect Four™ against each other

Connect Four™ is a popular two-player game sold under the Milton Bradley trademark. It is also known by other names such as Four in a Row or Four Up. In this game, the players take turns dropping discs into a vertical grid consisting of six rows and seven columns. The goal is to get four discs in a line. This is a variant of the Connect Four recipe given in the easyAI library. Let's see how to build it. In this recipe, instead of playing against the computer, we will create two bots that will play against each other. We will use a different algorithm for each to see which one wins.

Create a new Python file and import the following packages:

```
import numpy as np
from easyAI import TwoPlayersGame, Human_Player, AI_Player, \
        Negamax, SSS
```

Define a class that contains all the methods needed to play the game:

```
class GameController(TwoPlayersGame):
    def __init__(self, players, board = None):
        # Define the players
        self.players = players
```

Define the board with six rows and seven columns:

```
        # Define the configuration of the board
        self.board = board if (board != None) else (
            np.array([[0 for i in range(7)] for j in range(6)]))
```

Define who's going to start the game. In this case, let's have player one start the game:

```
        # Define who starts the game
        self.nplayer = 1
```

Define the positions:

```
        # Define the positions
        self.pos_dir = np.array([[[i, 0], [0, 1]] for i in range(6)] +
                    [[[0, i], [1, 0]] for i in range(7)] +
                    [[[i, 0], [1, 1]] for i in range(1, 3)] +
                    [[[0, i], [1, 1]] for i in range(4)] +
                    [[[i, 6], [1, -1]] for i in range(1, 3)] +
                    [[[0, i], [1, -1]] for i in range(3, 7)])
```

Define a method to get all the possible moves:

```
# Define possible moves
def possible_moves(self):
    return [i for i in range(7) if (self.board[:, i].min() == 0)]
```

Define a method to control how to make a move:

```
# Define how to make the move
def make_move(self, column):
    line = np.argmin(self.board[:, column] != 0)
    self.board[line, column] = self.nplayer
```

Define a method to show the current status:

```
# Show the current status
def show(self):
    print('\n' + '\n'.join(
            ['0 1 2 3 4 5 6', 13 * '-'] +
            [' '.join([['.', 'O', 'X'][self.board[5 - j][i]]
            for i in range(7)]) for j in range(6)]))
```

Define a method to compute what a loss looks like. Whenever somebody gets four in a line, that player wins the game:

```
# Define what a loss_condition looks like
def loss_condition(self):
    for pos, direction in self.pos_dir:
        streak = 0
        while (0 <= pos[0] <= 5) and (0 <= pos[1] <= 6):
            if self.board[pos[0], pos[1]] == self.nopponent:
                streak += 1
                if streak == 4:
                    return True
            else:
                streak = 0

            pos = pos + direction

    return False
```

Check if the game is over by using the `loss_condition` method:

```
# Check if the game is over
def is_over(self):
    return (self.board.min() > 0) or self.loss_condition()
```

Compute the score:

```
# Compute the score
def scoring(self):
    return -100 if self.loss_condition() else 0
```

Define the main function and start by defining the algorithms. We will let two algorithms play against each other. We will use Negamax for the first computer player and **SSS*** algorithm for the second computer player. SSS* is basically a search algorithm that conducts a state space search by traversing the tree in a best-first style. Both methods take, as an input argument, the number of turns in advance to think about. In this case, let's use five for both:

```
if __name__ == '__main__':
    # Define the algorithms that will be used
    algo_neg = Negamax(5)
    algo_sss = SSS(5)
```

Start playing the game:

```
# Start the game
game = GameController([AI_Player(algo_neg), AI_Player(algo_sss)])
game.play()
```

Print the result:

```
# Print the result
if game.loss_condition():
    print('\nPlayer', game.nopponent, 'wins. ')
else:
    print("\nIt's a draw.")
```

The full code is given in the file `connect_four.py`. This is not an interactive game. We are just pitting one algorithm against another. Negamax algorithm is player one and SSS* algorithm is player two.

If you run the code, you will get the following output on your Terminal at the beginning:

```
0 1 2 3 4 5 6
-------------
. . . . . . .
. . . . . . .
. . . . . . .
. . . . . . .
. . . . . . .
. . . . . . .
Move #1: player 1 plays 0 :

0 1 2 3 4 5 6
-------------
. . . . . . .
. . . . . . .
. . . . . . .
. . . . . . .
. . . . . . .
O . . . . . .
Move #2: player 2 plays 0 :

0 1 2 3 4 5 6
-------------
. . . . . . .
```

If you scroll down, you will see the following towards the end:

```
O O O X O O .
Move #35: player 1 plays 6 :

0 1 2 3 4 5 6
-------------
X X O O X . .
O O X X O . .
X X O O X X .
O O X X O O .
X X O X X X .
O O O X O O O
Move #36: player 2 plays 6 :

0 1 2 3 4 5 6
-------------
X X O O X . .
O O X X O . .
X X O O X X .
O O X X O O .
X X O X X X X
O O O X O O O
Player 2 wins.
```

As we can see, player two wins the game.

[247]

Building two bots to play Hexapawn against each other

Hexapawn is a two-player game where we start with a chessboard of size *NxM*. We have pawns on each side of the board and the goal is to advance a pawn all the way to the other end of the board. The standard pawn rules of chess are applicable here. This is a variant of the Hexapawn recipe given in the `easyAI` library. We will create two bots and pit an algorithm against itself to see what happens.

Create a new Python file and import the following packages:

```
from easyAI import TwoPlayersGame, AI_Player, \
        Human_Player, Negamax
```

Define a class that contains all the methods necessary to control the game. Start by defining the number of pawns on each side and the length of the board. Create a list of tuples containing the positions:

```
class GameController(TwoPlayersGame):
    def __init__(self, players, size = (4, 4)):
        self.size = size
        num_pawns, len_board = size
        p = [[(i, j) for j in range(len_board)] \
                for i in [0, num_pawns - 1]]
```

Assign the direction, goals, and pawns to each player:

```
        for i, d, goal, pawns in [(0, 1, num_pawns - 1,
                p[0]), (1, -1, 0, p[1])]:
            players[i].direction = d
            players[i].goal_line = goal
            players[i].pawns = pawns
```

Define the players and specify who starts first:

```
        # Define the players
        self.players = players

        # Define who starts first
        self.nplayer = 1
```

Define the alphabets that will be used to identify positions like B6 or C7 on a chessboard:

```
        # Define the alphabets
        self.alphabets = 'ABCDEFGHIJ'
```

[248]

Define a lambda function to convert strings to tuples:

```
# Convert B4 to (1, 3)
self.to_tuple = lambda s: (self.alphabets.index(s[0]),
        int(s[1:]) - 1)
```

Define a lambda function to convert tuples to strings:

```
# Convert (1, 3) to B4
self.to_string = lambda move: ' '.join([self.alphabets[
        move[i][0]] + str(move[i][1] + 1)
        for i in (0, 1)])
```

Define a method to compute the possible moves:

```
# Define the possible moves
def possible_moves(self):
    moves = []
    opponent_pawns = self.opponent.pawns
    d = self.player.direction
```

If you don't find an opponent pawn in a position, then that's a valid move:

```
    for i, j in self.player.pawns:
        if (i + d, j) not in opponent_pawns:
            moves.append(((i, j), (i + d, j)))

        if (i + d, j + 1) in opponent_pawns:
            moves.append(((i, j), (i + d, j + 1)))

        if (i + d, j - 1) in opponent_pawns:
            moves.append(((i, j), (i + d, j - 1)))

    return list(map(self.to_string, [(i, j) for i, j in moves]))
```

Define how to make a move and update the pawns based on that:

```
# Define how to make a move
def make_move(self, move):
    move = list(map(self.to_tuple, move.split(' ')))
    ind = self.player.pawns.index(move[0])
    self.player.pawns[ind] = move[1]

    if move[1] in self.opponent.pawns:
        self.opponent.pawns.remove(move[1])
```

Define the conditions for a loss. If a player gets 4 in a line, then the opponent loses:

```
# Define what a loss looks like
```

[249]

```python
def loss_condition(self):
    return (any([i == self.opponent.goal_line
                 for i, j in self.opponent.pawns])
            or (self.possible_moves() == []) )
```

Check if the game is over using the `loss_condition` method:

```python
# Check if the game is over
def is_over(self):
    return self.loss_condition()
```

Print the current status:

```python
# Show the current status
def show(self):
    f = lambda x: '1' if x in self.players[0].pawns else (
            '2' if x in self.players[1].pawns else '.')

    print("\n".join([" ".join([f((i, j))
            for j in range(self.size[1])])
            for i in range(self.size[0])]))
```

Define the main function and start by defining the scoring lambda function:

```python
if __name__=='__main__':
    # Compute the score
    scoring = lambda game: -100 if game.loss_condition() else 0
```

Define the algorithm to be used. In this case, we will use Negamax that can calculate 12 moves in advance and uses a `scoring` lambda function for strategy:

```python
    # Define the algorithm
    algorithm = Negamax(12, scoring)
```

Start playing the game:

```python
    # Start the game
    game = GameController([AI_Player(algorithm),
            AI_Player(algorithm)])
    game.play()
    print('\nPlayer', game.nopponent, 'wins after', game.nmove, 'turns')
```

The full code is given in the file `hexapawn.py`. It's not an interactive game. We are pitting an AI algorithm against itself. If you run the code, you will get the following output on your Terminal at the beginning:

```
1 1 1 1
. . . .
. . . .
2 2 2 2

Move #1: player 1 plays A1 B1 :
. 1 1 1
1 . . .
. . . .
2 2 2 2

Move #2: player 2 plays D1 C1 :
. 1 1 1
1 . . .
2 . . .
. 2 2 2

Move #3: player 1 plays A2 B2 :
. . 1 1
1 1 . .
2 . . .
. 2 2 2

Move #4: player 2 plays D2 C2 :
. . 1 1
```

If you scroll down, you will see the following towards the end:

```
Move #4: player 2 plays D2 C2 :
. . 1 1
1 1 . .
2 2 . .
. . 2 2

Move #5: player 1 plays B1 C2 :
. . 1 1
. 1 . .
2 1 . .
. . 2 2

Move #6: player 2 plays C1 B1 :
. . 1 1
2 1 . .
. 1 . .
. . 2 2

Move #7: player 1 plays C2 D2 :
. . 1 1
2 1 . .
. . . .
. 1 2 2

Player 1 wins after 8 turns
```

As we can see, player one wins the game.

Summary

In this chapter, we discussed how to build games with artificial intelligence, and how to use search algorithms to effectively come up with strategies to win the games. We talked about combinatorial search and how it can be used to speed up the search process. We learned about Minimax and Alpha-Beta pruning. We learned how the Negamax algorithm is used in practice. We then used these algorithms to build bots to play Last Coin Standing and Tic-Tac-Toe.

We learned how to build two bots to play against each other in Connect Four and Hexapawn. In the next chapter, we will discuss natural language processing and how to use it to analyze text data by modeling and classifying it.

10
Natural Language Processing

In this chapter, we are going to learn about natural language processing. We will discuss various concepts such as tokenization, stemming, and lemmatization to process text. We will then discuss how to build a Bag of Words model and use it to classify text. We will see how to use machine learning to analyze the sentiment of a given sentence. We will then discuss topic modeling and implement a system to identify topics in a given document.

By the end of this chapter, you will know:

- How to install relevant packages
- Tokenizing text data
- Converting words to their base forms using stemming
- Converting words to their base forms using lemmatization
- Dividing text data into chunks
- Extracting document term matrix using the Bag of Words model
- Building a category predictor
- Constructing a gender identifier
- Building a sentiment analyzer
- Topic modeling using Latent Dirichlet Allocation

Introduction and installation of packages

Natural Language Processing (**NLP**) has become an important part of modern systems. It is used extensively in search engines, conversational interfaces, document processors, and so on. Machines can handle structured data well. But when it comes to working with free-form text, they have a hard time. The goal of NLP is to develop algorithms that enable computers to understand freeform text and help them understand language.

One of the most challenging things about processing freeform natural language is the sheer number of variations. The context plays a very important role in how a particular sentence is understood. Humans are great at these things because we have been trained for many years. We immediately use our past knowledge to understand the context and know what the other person is talking about.

To address this issue, NLP researchers started developing various applications using machine learning approaches. To build such applications, we need to collect a large corpus of text and then train the algorithm to perform various tasks like categorizing text, analyzing sentiments, or modeling topics. These algorithms are trained to detect patterns in input text data and derive insights from it.

In this chapter, we will discuss various underlying concepts that are used to analyze text and build NLP applications. This will enable us to understand how to extract meaningful information from the given text data. We will use a Python package called **Natural Language Toolkit** (**NLTK**) to build these applications. Make sure that you install this before you proceed. You can install it by running the following command on your Terminal:

```
$ pip3 install nltk
```

You can find more information about NLTK at http://www.nltk.org.

In order to access all the datasets provided by NLTK, we need to download it. Open up a Python shell by typing the following on your Terminal:

```
$ python3
```

We are now inside the Python shell. Type the following to download the data:

```
>>> import nltk
>>> nltk.download()
```

We will also use a package called `gensim` in this chapter. It's a robust semantic modeling library that's useful for many applications. You can install it by running the following command on your Terminal:

```
$ pip3 install gensim
```

You might need another package called `pattern` for `gensim` to function properly. You can install it by running the following command on your Terminal:

```
$ pip3 install pattern
```

You can find more information about `gensim` at https://radimrehurek.com/gensim. Now that you have installed the `NLTK` and `gensim`, let's proceed with the discussion.

Tokenizing text data

When we deal with text, we need to break it down into smaller pieces for analysis. This is where tokenization comes into the picture. It is the process of dividing the input text into a set of pieces like words or sentences. These pieces are called tokens. Depending on what we want to do, we can define our own methods to divide the text into many tokens. Let's take a look at how to tokenize the input text using `NLTK`.

Create a new Python file and import the following packages:

```
from nltk.tokenize import sent_tokenize, \
        word_tokenize, WordPunctTokenizer
```

Define some input text that will be used for tokenization:

```
# Define input text
input_text = "Do you know how tokenization works? It's actually quite interesting! Let's analyze a couple of sentences and figure it out."
```

Divide the input text into sentence tokens:

```
# Sentence tokenizer
print("\nSentence tokenizer:")
print(sent_tokenize(input_text))
```

Divide the input text into word tokens:

```
# Word tokenizer
print("\nWord tokenizer:")
print(word_tokenize(input_text))
```

Divide the input text into word tokens using word punct tokenizer:

```
# WordPunct tokenizer
print("\nWord punct tokenizer:")
print(WordPunctTokenizer().tokenize(input_text))
```

The full code is given in the file `tokenizer.py`. If you run the code, you will get the following output on your Terminal:

```
Sentence tokenizer:
['Do you know how tokenization works?', "It's actually quite interesting!", "Let's analyze a couple of se
ntences and figure it out."]
Word tokenizer:
['Do', 'you', 'know', 'how', 'tokenization', 'works', '?', 'It', "'s", 'actually', 'quite', 'interesting'
, '!', 'Let', "'s", 'analyze', 'a', 'couple', 'of', 'sentences', 'and', 'figure', 'it', 'out', '.']
Word punct tokenizer:
['Do', 'you', 'know', 'how', 'tokenization', 'works', '?', 'It', "'", 's', 'actually', 'quite', 'interest
ing', '!', 'Let', "'", 's', 'analyze', 'a', 'couple', 'of', 'sentences', 'and', 'figure', 'it', 'out', '.
']
```

We can see that the sentence tokenizer divides the input text into sentences. The two word tokenizers behave differently when it comes to punctuation. For example, the word "It's" is divided differently in the punct tokenizer as compared to the regular tokenizer.

Converting words to their base forms using stemming

Working with text has a lot of variations included in it. We have to deal with different forms of the same word and enable the computer to understand that these different words have the same base form. For example, the word *sing* can appear in many forms such as *sang, singer, singing, singer*, and so on. We just saw a set of words with similar meanings. Humans can easily identify these base forms and derive context.

When we analyze text, it's useful to extract these base forms. It will enable us to extract useful statistics to analyze the input text. Stemming is one way to achieve this. The goal of a stemmer is to reduce words in their different forms into a common base form. It is basically a heuristic process that cuts off the ends of words to extract their base forms. Let's see how to do it using NLTK.

Create a new python file and import the following packages:

```
from nltk.stem.porter import PorterStemmer
from nltk.stem.lancaster import LancasterStemmer
from nltk.stem.snowball import SnowballStemmer
```

Define some input words:

```
input_words = ['writing', 'calves', 'be', 'branded', 'horse', 'randomize',
        'possibly', 'provision', 'hospital', 'kept', 'scratchy', 'code']
```

Create objects for **Porter**, **Lancaster**, and **Snowball** stemmers:

```
# Create various stemmer objects
porter = PorterStemmer()
lancaster = LancasterStemmer()
snowball = SnowballStemmer('english')
```

Create a list of names for table display and format the output text accordingly:

```
#Create a list of stemmer names for display
stemmer_names = ['PORTER', 'LANCASTER', 'SNOWBALL']
formatted_text = '{:>16}' * (len(stemmer_names) + 1)
print('\n', formatted_text.format('INPUT WORD', *stemmer_names),
        '\n', '='*68)
```

Iterate through the words and stem them using the three stemmers:

```
# Stem each word and display the output
for word in input_words:
    output = [word, porter.stem(word),
            lancaster.stem(word), snowball.stem(word)]
    print(formatted_text.format(*output))
```

The full code is given in the file `stemmer.py`. If you run the code, you will get the following output on your Terminal:

INPUT WORD	PORTER	LANCASTER	SNOWBALL
writing	write	writ	write
calves	calv	calv	calv
be	be	be	be
branded	brand	brand	brand
horse	hors	hors	hors
randomize	random	random	random
possibly	possibl	poss	possibl
provision	provis	provid	provis
hospital	hospit	hospit	hospit
kept	kept	kept	kept
scratchy	scratchi	scratchy	scratchi
code	code	cod	code

Let's talk a bit about the three stemming algorithms that are being used here. All of them basically try to achieve the same goal. The difference between them is the level of strictness that's used to arrive at the base form.

The Porter stemmer is the least in terms of strictness and Lancaster is the strictest. If you closely observe the outputs, you will notice the differences. Stemmers behave differently when it comes to words like `possibly` or `provision`. The stemmed outputs that are obtained from the Lancaster stemmer are a bit obfuscated because it reduces the words a lot. At the same time, the algorithm is really fast. A good rule of thumb is to use the Snowball stemmer because it's a good trade off between speed and strictness.

Converting words to their base forms using lemmatization

Lemmatization is another way of reducing words to their base forms. In the previous section, we saw that the base forms that were obtained from those stemmers didn't make sense. For example, all the three stemmers said that the base form of *calves* is *calv*, which is not a real word. Lemmatization takes a more structured approach to solve this problem.

The lemmatization process uses a vocabulary and morphological analysis of words. It obtains the base forms by removing the inflectional word endings such as *ing* or *ed*. This base form of any word is known as the lemma. If you lemmatize the word *calves*, you should get *calf* as the output. One thing to note is that the output depends on whether the word is a verb or a noun. Let's take a look at how to do this using NLTK.

Create a new python file and import the following packages:

```
from nltk.stem import WordNetLemmatizer
```

Define some input words. We will be using the same set of words that we used in the previous section so that we can compare the outputs.

```
input_words = ['writing', 'calves', 'be', 'branded', 'horse', 'randomize',
        'possibly', 'provision', 'hospital', 'kept', 'scratchy', 'code']
```

Create a `lemmatizer` object:

```
# Create lemmatizer object
lemmatizer = WordNetLemmatizer()
```

Create a list of `lemmatizer` names for table display and format the text accordingly:

```
# Create a list of lemmatizer names for display
lemmatizer_names = ['NOUN LEMMATIZER', 'VERB LEMMATIZER']
formatted_text = '{:>24}' * (len(lemmatizer_names) + 1)
print('\n', formatted_text.format('INPUT WORD', *lemmatizer_names),
      '\n', '='*75)
```

Iterate through the words and lemmatize the words using Noun and Verb lemmatizers:

```
# Lemmatize each word and display the output
for word in input_words:
    output = [word, lemmatizer.lemmatize(word, pos='n'),
              lemmatizer.lemmatize(word, pos='v')]
    print(formatted_text.format(*output))
```

The full code is given in the file `lemmatizer.py`. If you run the code, you will get the following output on your Terminal:

INPUT WORD	NOUN LEMMATIZER	VERB LEMMATIZER
writing	writing	write
calves	calf	calve
be	be	be
branded	branded	brand
horse	horse	horse
randomize	randomize	randomize
possibly	possibly	possibly
provision	provision	provision
hospital	hospital	hospital
kept	kept	keep
scratchy	scratchy	scratchy
code	code	code

We can see that the noun `lemmatizer` works differently than the verb `lemmatizer` when it comes to words like *writing* or *calves*. If you compare these outputs to stemmer outputs, you will see that there are differences too. The `lemmatizer` outputs are all meaningful whereas stemmer outputs may or may not be meaningful.

Dividing text data into chunks

Text data usually needs to be divided into pieces for further analysis. This process is known as chunking. This is used frequently in text analysis. The conditions that are used to divide the text into chunks can vary based on the problem at hand. This is not the same as tokenization where we also divide text into pieces. During chunking, we do not adhere to any constraints and the output chunks need to be meaningful.

When we deal with large text documents, it becomes important to divide the text into chunks to extract meaningful information. In this section, we will see how to divide the input text into a number of pieces.

Create a new python file and import the following packages:

```
import numpy as np
from nltk.corpus import brown
```

Define a function to divide the input text into chunks. The first parameter is the text and the second parameter is the number of words in each chunk:

```
# Split the input text into chunks, where
# each chunk contains N words
def chunker(input_data, N):
    input_words = input_data.split(' ')
    output = []
```

Iterate through the words and divide them into chunks using the input parameter. The function returns a list:

```
    cur_chunk = []
    count = 0
    for word in input_words:
        cur_chunk.append(word)
        count += 1
        if count == N:
            output.append(' '.join(cur_chunk))
            count, cur_chunk = 0, []

    output.append(' '.join(cur_chunk))

    return output
```

Define the main function and read the input data using the Brown corpus. We will read `12,000` words in this case. You are free to read as many words as you want.

```
if __name__=='__main__':
    # Read the first 12000 words from the Brown corpus
    input_data = ' '.join(brown.words()[:12000])
```

Define the number of words in each chunk:

```
    # Define the number of words in each chunk
    chunk_size = 700
```

Divide the input text into chunks and display the output:

```
    chunks = chunker(input_data, chunk_size)
    print('\nNumber of text chunks =', len(chunks), '\n')
    for i, chunk in enumerate(chunks):
        print('Chunk', i+1, '==>', chunk[:50])
```

The full code is given in the file `text_chunker.py`. If you run the code, you will get the following output on your Terminal:

```
Number of text chunks = 18

Chunk 1 ==> The Fulton County Grand Jury said Friday an invest
Chunk 2 ==> '' . ( 2 ) Fulton legislators `` work with city of
Chunk 3 ==> . Construction bonds Meanwhile , it was learned th
Chunk 4 ==> , anonymous midnight phone calls and veiled threat
Chunk 5 ==> Harris , Bexar , Tarrant and El Paso would be $451
Chunk 6 ==> set it for public hearing on Feb. 22 . The proposa
Chunk 7 ==> College . He has served as a border patrolman and
Chunk 8 ==> of his staff were doing on the address involved co
Chunk 9 ==> plan alone would boost the base to $5,000 a year a
Chunk 10 ==> nursing homes In the area of `` community health s
Chunk 11 ==> of its Angola policy prove harsh , there has been
Chunk 12 ==> system which will prevent Laos from being used as
Chunk 13 ==> reform in recipient nations . In Laos , the admini
Chunk 14 ==> . He is not interested in being named a full-time
Chunk 15 ==> said , `` to obtain the views of the general publi
Chunk 16 ==> '' . Mr. Reama , far from really being retired , i
Chunk 17 ==> making enforcement of minor offenses more effectiv
Chunk 18 ==> to tell the people where he stands on the tax issu
```

The preceding screenshot shows the first 50 characters of each chunk.

Extracting the frequency of terms using a Bag of Words model

One of the main goals of text analysis is to convert text into numeric form so that we can use machine learning on it. Let's consider text documents that contain many millions of words. In order to analyze these documents, we need to extract the text and convert it into a form of numeric representation.

Machine learning algorithms need numeric data to work with so that they can analyze the data and extract meaningful information. This is where the Bag of Words model comes into picture. This model extracts a vocabulary from all the words in the documents and builds a model using a document term matrix. This allows us to represent every document as a *bag of words*. We just keep track of word counts and disregard the grammatical details and the word order.

Let's see what a document-term matrix is all about. A document term matrix is basically a table that gives us counts of various words that occur in the document. So a text document can be represented as a weighted combination of various words. We can set thresholds and choose words that are more meaningful. In a way, we are building a histogram of all the words in the document that will be used as a feature vector. This feature vector is used for text classification.

Consider the following sentences:

- Sentence 1: The children are playing in the hall
- Sentence 2: The hall has a lot of space
- Sentence 3: Lots of children like playing in an open space

If you consider all the three sentences, we have the following nine unique words:

- the
- children
- are
- playing
- in
- hall
- has

- a
- lot
- of
- space
- like
- an
- open

There are 14 distinct words here. Let's construct a histogram for each sentence by using the word count in each sentence. Each feature vector will be 14-dimensional because we have 14 distinct words overall:

- Sentence 1: [2, 1, 1, 1, 1, 1, 0, 0, 0, 0, 0, 0, 0, 0]
- Sentence 2: [1, 0, 0, 0, 0, 1, 1, 1, 1, 1, 1, 0, 0, 0]
- Sentence 3: [0, 1, 0, 1, 1, 0, 0, 0, 1, 1, 1, 1, 1, 1]

Now that we have extracted these feature vectors, we can use machine learning algorithms to analyze this data.

Let's see how to build a Bag of Words model in NLTK. Create a new python file and import the following packages:

```
import numpy as np
from sklearn.feature_extraction.text import CountVectorizer
from nltk.corpus import brown
from text_chunker import chunker
```

Read the input data from Brown corpus. We will read 5,400 words. You are free to read as many number of words as you want.

```
# Read the data from the Brown corpus
input_data = ' '.join(brown.words()[:5400])
```

Define the number of words in each chunk:

```
# Number of words in each chunk
chunk_size = 800
```

Divide the input text into chunks:

```
text_chunks = chunker(input_data, chunk_size)
```

Convert the chunks into dictionary items:

```
# Convert to dict items
chunks = []
for count, chunk in enumerate(text_chunks):
    d = {'index': count, 'text': chunk}
    chunks.append(d)
```

Extract the document term matrix where we get the count of each word. We will achieve this using the `CountVectorizer` method that takes two input parameters. The first parameter is the minimum document frequency and the second parameter is the maximum document frequency. The frequency refers to the number of occurrences of a word in the text.

```
# Extract the document term matrix
count_vectorizer = CountVectorizer(min_df=7, max_df=20)
document_term_matrix = count_vectorizer.fit_transform([chunk['text'] for chunk in chunks])
```

Extract the vocabulary and display it. The vocabulary refers to the list of distinct words that were extracted in the previous step.

```
# Extract the vocabulary and display it
vocabulary = np.array(count_vectorizer.get_feature_names())
print("\nVocabulary:\n", vocabulary)
```

Generate the names for display:

```
# Generate names for chunks
chunk_names = []
for i in range(len(text_chunks)):
    chunk_names.append('Chunk-' + str(i+1))
```

Print the document term matrix:

```
# Print the document term matrix
print("\nDocument term matrix:")
formatted_text = '{:>12}' * (len(chunk_names) + 1)
print('\n', formatted_text.format('Word', *chunk_names), '\n')
for word, item in zip(vocabulary, document_term_matrix.T):
    # 'item' is a 'csr_matrix' data structure
    output = [word] + [str(freq) for freq in item.data]
    print(formatted_text.format(*output))
```

The full code is given in the file `bag_of_words.py`. If you run the code, you will get the following output on your Terminal:

```
Document term matrix:
      Word    Chunk-1    Chunk-2    Chunk-3    Chunk-4    Chunk-5    Chunk-6    Chunk-7
       and         23          9          9         11          9         17         10
       are          2          2          1          1          2          2          1
        be          6          8          7          7          6          2          1
        by          3          4          4          5         14          3          6
    county          6          2          7          3          1          2          2
       for          7         13          4         10          7          6          4
        in         15         11         15         11         13         14         17
        is          2          7          3          4          5          5          2
        it          8          6          8          9          3          1          2
        of         31         20         20         30         29         35         26
        on          4          3          5         10          6          5          2
       one          1          3          1          2          2          1          1
      said         12          5          7          7          4          3          7
     state          3          7          2          6          3          4          1
      that         13          8          9          2          7          1          7
       the         71         51         43         51         43         52         49
        to         11         26         20         26         21         15         11
       two          2          1          1          1          1          2          2
       was          5          6          7          7          4          7          3
     which          7          4          5          4          3          1          1
      with          2          2          3          1          2          2          3
```

We can see all the words in the document term matrix and the corresponding counts in each chunk.

Building a category predictor

A category predictor is used to predict the category to which a given piece of text belongs. This is frequently used in text classification to categorize text documents. Search engines frequently use this tool to order the search results by relevance. For example, let's say that we want to predict whether a given sentence belongs to sports, politics, or science. To do this, we build a corpus of data and train an algorithm. This algorithm can then be used for inference on unknown data.

In order to build this predictor, we will use a statistic called **TermFrequency – Inverse Document Frequency (tf-idf)**. In a set of documents, we need to understand the importance of each word. The tf-idf statistic helps us understand how important a given word is to a document in a set of documents.

Let's consider the first part of this statistic. The **Term Frequency** (**tf**) is basically a measure of how frequently each word appears in a given document. Since different documents have a different number of words, the exact numbers in the histogram will vary. In order to have a level playing field, we need to normalize the histograms. So we divide the count of each word by the total number of words in a given document to obtain the term frequency.

The second part of the statistic is the **Inverse Document Frequency** (**idf**), which is a measure of how unique a word is to this document in the given set of documents. When we compute the term frequency, the assumption is that all the words are equally important. But we cannot just rely on the frequency of each word because words like *and* and *the* appear a lot. To balance the frequencies of these commonly occurring words, we need to reduce their weights and weigh up the rare words. This helps us identify words that are unique to each document as well, which in turn helps us formulate a distinctive feature vector.

To compute this statistic, we need to compute the ratio of the number of documents with the given word and divide it by the total number of documents. This ratio is essentially the fraction of the documents that contain the given word. Inverse document frequency is then calculated by taking the negative algorithm of this ratio.

We then combine term frequency and inverse document frequency to formulate a feature vector to categorize documents. Let's see how to build a category predictor.

Create a new python file and import the following packages:

```
from sklearn.datasets import fetch_20newsgroups
from sklearn.naive_bayes import MultinomialNB
from sklearn.feature_extraction.text import TfidfTransformer
from sklearn.feature_extraction.text import CountVectorizer
```

Define the map of categories that will be used for training. We will be using five categories in this case. The keys in this dictionary object refer to the names in the `scikit-learn` dataset.

```
# Define the category map
category_map = {'talk.politics.misc': 'Politics', 'rec.autos': 'Autos',
        'rec.sport.hockey': 'Hockey', 'sci.electronics': 'Electronics',
        'sci.med': 'Medicine'}
```

Get the training dataset using `fetch_20newsgroups`:

```
# Get the training dataset
training_data = fetch_20newsgroups(subset='train',
        categories=category_map.keys(), shuffle=True, random_state=5)
```

Extract the term counts using the `CountVectorizer` object:

```
# Build a count vectorizer and extract term counts
count_vectorizer = CountVectorizer()
train_tc = count_vectorizer.fit_transform(training_data.data)
print("\nDimensions of training data:", train_tc.shape)
```

Create Term Frequency – Inverse Document Frequency (tf-idf) transformer and train it using the data:

```
# Create the tf-idf transformer
tfidf = TfidfTransformer()
train_tfidf = tfidf.fit_transform(train_tc)
```

Define some sample input sentences that will be used for testing:

```
# Define test data
input_data = [
    'You need to be careful with cars when you are driving on slippery roads',
    'A lot of devices can be operated wirelessly',
    'Players need to be careful when they are close to goal posts',
    'Political debates help us understand the perspectives of both sides'
]
```

Train a Multinomial Bayes classifier using the training data:

```
# Train a Multinomial Naive Bayes classifier
classifier = MultinomialNB().fit(train_tfidf, training_data.target)
```

Transform the input data using the count `vectorizer`:

```
# Transform input data using count vectorizer
input_tc = count_vectorizer.transform(input_data)
```

Transform the vectorized data using the `tf-idf` transformer so that it can be run through the inference model:

```
# Transform vectorized data using tfidf transformer
input_tfidf = tfidf.transform(input_tc)
```

Predict the output using the `tf-idf` transformed vector:

```
# Predict the output categories
predictions = classifier.predict(input_tfidf)
```

Print the output category for each sample in the input test data:

```
# Print the outputs
for sent, category in zip(input_data, predictions):
    print('\nInput:', sent, '\nPredicted category:', \
            category_map[training_data.target_names[category]])
```

The full code is given in the file `category_predictor.py`. If you run the code, you will get the following output on your Terminal:

```
Dimensions of training data: (2844, 40321)

Input: You need to be careful with cars when you are driving on slippery roads
Predicted category: Autos

Input: A lot of devices can be operated wirelessly
Predicted category: Electronics

Input: Players need to be careful when they are close to goal posts
Predicted category: Hockey

Input: Political debates help us understand the perspectives of both sides
Predicted category: Politics
```

We can see intuitively that the predicted categories are correct.

Constructing a gender identifier

Gender identification is an interesting problem. In this case, we will use the heuristic to construct a feature vector and use it to train a classifier. The heuristic that will be used here is the last *N* letters of a given name. For example, if the name ends with *ia*, it's most likely a female name, such as *Amelia* or *Genelia*. On the other hand, if the name ends with *rk*, it's likely a male name such as *Mark* or *Clark*. Since we are not sure of the exact number of letters to use, we will play around with this parameter and find out what the best answer is. Let's see how to do it.

Create a new python file and import the following packages:

```
import random

from nltk import NaiveBayesClassifier
from nltk.classify import accuracy as nltk_accuracy
from nltk.corpus import names
```

Define a function to extract the last *N* letters from the input word:

```
# Extract last N letters from the input word
# and that will act as our "feature"
def extract_features(word, N=2):
    last_n_letters = word[-N:]
    return {'feature': last_n_letters.lower()}
```

Define the main function and extract training data from the `scikit-learn` package. This data contains labeled male and female names:

```
if __name__=='__main__':
    # Create training data using labeled names available in NLTK
    male_list = [(name, 'male') for name in names.words('male.txt')]
    female_list = [(name, 'female') for name in names.words('female.txt')]
    data = (male_list + female_list)
```

Seed the random number generator and shuffle the data:

```
    # Seed the random number generator
    random.seed(5)

    # Shuffle the data
    random.shuffle(data)
```

Create some sample names that will be used for testing:

```
    # Create test data
    input_names = ['Alexander', 'Danielle', 'David', 'Cheryl']
```

Define the percentage of data that will be used for training and testing:

```
    # Define the number of samples used for train and test
    num_train = int(0.8 * len(data))
```

We will be using the last *N* characters as the feature vector to predict the gender. We will vary this parameter to see how the performance varies. In this case, we will go from 1 to 6:

```
    # Iterate through different lengths to compare the accuracy
    for i in range(1, 6):
        print('\nNumber of end letters:', i)
        features = [(extract_features(n, i), gender) for (n, gender) in data]
```

Separate the data into training and testing:

```
        train_data, test_data = features[:num_train], features[num_train:]
```

Build a `NaiveBayes` Classifier using the training data:

```
classifier = NaiveBayesClassifier.train(train_data)
```

Compute the accuracy of the classifier using the inbuilt method available in `NLTK`:

```
# Compute the accuracy of the classifier
accuracy = round(100 * nltk_accuracy(classifier, test_data), 2)
print('Accuracy = ' + str(accuracy) + '%')
```

Predict the output for each name in the input test list:

```
# Predict outputs for input names using the trained classifier model
for name in input_names:
    print(name, '==>', classifier.classify(extract_features(name, i)))
```

The full code is given in the file `gender_identifier.py`. If you run the code, you will get the following output on your Terminal:

```
Number of end letters: 1
Accuracy = 74.7%
Alexander ==> male
Danielle ==> female
David ==> male
Cheryl ==> male

Number of end letters: 2
Accuracy = 78.79%
Alexander ==> male
Danielle ==> female
David ==> male
Cheryl ==> female

Number of end letters: 3
Accuracy = 77.22%
Alexander ==> male
Danielle ==> female
David ==> male
Cheryl ==> female
```

The preceding screenshot shows the accuracy as well as the predicted outputs for the test data. Let's go further and see what happens:

```
Number of end letters: 4
Accuracy = 69.98%
Alexander ==> male
Danielle ==> female
David ==> male
Cheryl ==> female

Number of end letters: 5
Accuracy = 64.63%
Alexander ==> male
Danielle ==> female
David ==> male
Cheryl ==> female
```

We can see that the accuracy peaked at two letters and then started decreasing after that.

Building a sentiment analyzer

Sentiment analysis is the process of determining the sentiment of a given piece of text. For example, it can used to determine whether a movie review is positive or negative. This is one of the most popular applications of natural language processing. We can add more categories as well depending on the problem at hand. This technique is generally used to get a sense of how people feel about a particular product, brand, or topic. It is frequently used to analyze marketing campaigns, opinion polls, social media presence, product reviews on e-commerce sites, and so on. Let's see how to determine the sentiment of a movie review.

We will use a Naive Bayes classifier to build this classifier. We first need to extract all the unique words from the text. The NLTK classifier needs this data to be arranged in the form of a dictionary so that it can ingest it. Once we divide the text data into training and testing datasets, we will train the Naive Bayes classifier to classify the reviews into positive and negative. We will also print out the top informative words to indicate positive and negative reviews. This information is interesting because it tells us what words are being used to denote various reactions.

Create a new python file and import the following packages:

```
from nltk.corpus import movie_reviews
from nltk.classify import NaiveBayesClassifier
from nltk.classify.util import accuracy as nltk_accuracy
```

Define a function to construct a dictionary object based on the input words and return it:

```
# Extract features from the input list of words
def extract_features(words):
    return dict([(word, True) for word in words])
```

Define the main function and load the labeled movie reviews:

```
if __name__=='__main__':
    # Load the reviews from the corpus
    fileids_pos = movie_reviews.fileids('pos')
    fileids_neg = movie_reviews.fileids('neg')
```

Extract the features from the movie reviews and label it accordingly:

```
    # Extract the features from the reviews
    features_pos = [(extract_features(movie_reviews.words(
            fileids=[f])), 'Positive') for f in fileids_pos]
    features_neg = [(extract_features(movie_reviews.words(
            fileids=[f])), 'Negative') for f in fileids_neg]
```

Define the split between training and testing. In this case, we will allocate 80% for training and 20% for testing:

```
    # Define the train and test split (80% and 20%)
    threshold = 0.8
    num_pos = int(threshold * len(features_pos))
    num_neg = int(threshold * len(features_neg))
```

Separate the feature vectors for training and testing:

```
    # Create training and training datasets
    features_train = features_pos[:num_pos] + features_neg[:num_neg]
    features_test = features_pos[num_pos:] + features_neg[num_neg:]
```

Print the number of datapoints used for training and testing:

```
    # Print the number of datapoints used
    print('\nNumber of training datapoints:', len(features_train))
    print('Number of test datapoints:', len(features_test))
```

Train a `Naive Bayes` classifier using the training data and compute the accuracy using the inbuilt method available in `NLTK`:

```
# Train a Naive Bayes classifier
classifier = NaiveBayesClassifier.train(features_train)
print('\nAccuracy of the classifier:', nltk_accuracy(
        classifier, features_test))
```

Print the top N most informative words:

```
N = 15
print('\nTop ' + str(N) + ' most informative words:')
for i, item in enumerate(classifier.most_informative_features()):
    print(str(i+1) + '. ' + item[0])
    if i == N - 1:
        break
```

Define sample sentences to be used for testing:

```
# Test input movie reviews
input_reviews = [
    'The costumes in this movie were great',
    'I think the story was terrible and the characters were very weak',
    'People say that the director of the movie is amazing',
    'This is such an idiotic movie. I will not recommend it to anyone.'
]
```

Iterate through the sample data and predict the output:

```
print("\nMovie review predictions:")
for review in input_reviews:
    print("\nReview:", review)
```

Compute the probabilities for each class:

```
    # Compute the probabilities
    probabilities = 
classifier.prob_classify(extract_features(review.split()))
```

Pick the maximum value among the probabilities:

```
    # Pick the maximum value
    predicted_sentiment = probabilities.max()
```

Print the predicted output class (positive or negative sentiment):

```
# Print outputs
print("Predicted sentiment:", predicted_sentiment)
print("Probability:", 
round(probabilities.prob(predicted_sentiment), 2))
```

The full code is given in the file `sentiment_analyzer.py`. If you run the code, you will get the following output on your Terminal:

```
Number of training datapoints: 1600
Number of test datapoints: 400

Accuracy of the classifier: 0.735

Top 15 most informative words:
1. outstanding
2. insulting
3. vulnerable
4. ludicrous
5. uninvolving
6. astounding
7. avoids
8. fascination
9. symbol
10. seagal
11. affecting
12. anna
13. darker
14. animators
15. idiotic
```

The preceding screenshot shows the top 15 most informative words. If you scroll down your Terminal, you will see this:

```
Movie review predictions:

Review: The costumes in this movie were great
Predicted sentiment: Positive
Probability: 0.59

Review: I think the story was terrible and the characters were very weak
Predicted sentiment: Negative
Probability: 0.8

Review: People say that the director of the movie is amazing
Predicted sentiment: Positive
Probability: 0.6

Review: This is such an idiotic movie. I will not recommend it to anyone.
Predicted sentiment: Negative
Probability: 0.87
```

We can see and verify intuitively that the predictions are correct.

Topic modeling using Latent Dirichlet Allocation

Topic modeling is the process of identifying patterns in text data that correspond to a topic. If the text contains multiple topics, then this technique can be used to identify and separate those themes within the input text. We do this to uncover hidden thematic structure in the given set of documents.

Topic modeling helps us to organize our documents in an optimal way, which can then be used for analysis. One thing to note about topic modeling algorithms is that we don't need any labeled data. It is like unsupervised learning where it will identify the patterns on its own. Given the enormous volumes of text data generated on the Internet, topic modeling becomes very important because it enables us to summarize all this data, which would otherwise not be possible.

Latent Dirichlet Allocation is a topic modeling technique where the underlying intuition is that a given piece of text is a combination of multiple topics. Let's consider the following sentence – *Data visualization is an important tool in financial analysis*. This sentence has multiple topics like data, visualization, finance, and so on. This particular combination helps us identify this text in a large document. In essence, it is a statistical model that tries to capture this idea and create a model based on it. The model assumes that documents are generated from a random process based on these topics. A *topic* is basically a distribution over a fixed vocabulary of words. Let's see how to do topic modeling in Python.

We will use a library called `gensim` in this section. We have already installed this library in the first section of this chapter. Make sure that you have it before you proceed. Create a new python file and import the following packages:

```
from nltk.tokenize import RegexpTokenizer
from nltk.corpus import stopwords
from nltk.stem.snowball import SnowballStemmer
from gensim import models, corpora
```

Define a function to load the input data. The input file contains 10 line-separated sentences:

```
# Load input data
def load_data(input_file):
    data = []
    with open(input_file, 'r') as f:
        for line in f.readlines():
            data.append(line[:-1])

    return data
```

Define a function to process the input text. The first step is to tokenize it:

```
# Processor function for tokenizing, removing stop
# words, and stemming
def process(input_text):
    # Create a regular expression tokenizer
    tokenizer = RegexpTokenizer(r'\w+')
```

We then need to stem the tokenized text:

```
# Create a Snowball stemmer
stemmer = SnowballStemmer('english')
```

We need to remove the stop words from the input text because they don't add information. Let's get the list of stop-words:

```
# Get the list of stop words
stop_words = stopwords.words('english')
```

Tokenize the input string:

```
# Tokenize the input string
tokens = tokenizer.tokenize(input_text.lower())
```

Remove the stop-words:

```
# Remove the stop words
tokens = [x for x in tokens if not x in stop_words]
```

Stem the tokenized words and return the list:

```
# Perform stemming on the tokenized words
tokens_stemmed = [stemmer.stem(x) for x in tokens]

return tokens_stemmed
```

Define the main function and load the input data from the file data.txt provided to you:

```
if __name__=='__main__':
    # Load input data
    data = load_data('data.txt')
```

Tokenize the text:

```
# Create a list for sentence tokens
tokens = [process(x) for x in data]
```

Create a dictionary based on the tokenized sentences:

```
# Create a dictionary based on the sentence tokens
dict_tokens = corpora.Dictionary(tokens)
```

Create a document term matrix using the sentence tokens:

```
# Create a document-term matrix
doc_term_mat = [dict_tokens.doc2bow(token) for token in tokens]
```

We need to provide the number of topics as the input parameter. In this case, we know that the input text has two distinct topics. Let's specify that.

```
# Define the number of topics for the LDA model
num_topics = 2
```

Generate the LatentDirichlet Model:

```
# Generate the LDA model
ldamodel = models.ldamodel.LdaModel(doc_term_mat,
        num_topics=num_topics, id2word=dict_tokens, passes=25)
```

Print the top 5 contributing words for each topic:

```
num_words = 5
print('\nTop ' + str(num_words) + ' contributing words to each topic:')
for item in ldamodel.print_topics(num_topics=num_topics, num_words=num_words):
        print('\nTopic', item[0])

        # Print the contributing words along with their relative contributions
        list_of_strings = item[1].split(' + ')
        for text in list_of_strings:
            weight = text.split('*')[0]
            word = text.split('*')[1]
            print(word, '==>', str(round(float(weight) * 100, 2)) + '%')
```

The full code is given in the file `topic_modeler.py`. If you run the code, you will get the following output on your Terminal:

```
Top 5 contributing words to each topic:

Topic 0
mathemat ==> 2.7%
structur ==> 2.6%
set ==> 2.6%
formul ==> 2.6%
tradit ==> 1.6%

Topic 1
empir ==> 4.7%
expand ==> 3.3%
time ==> 2.0%
peopl ==> 2.0%
histor ==> 2.0%
```

We can see that it does a reasonably good job of separating the two topics – mathematics and history. If you look into the text, you can verify that each sentence is either about mathematics or history.

Summary

In this chapter, we learned about the various underlying concepts in natural language processing. We discussed tokenization and how to separate input text into multiple tokens. We learned how to reduce words to their base forms using stemming and lemmatization. We implemented a text chunker to divide input text into chunks based on predefined conditions.

We discussed the Bag of Words model and built a document term matrix for input text. We then learnt how to categorize text using machine learning. We constructed a gender identifier using a heuristic. We used machine learning to analyze the sentiments of movie reviews. We discussed topic modeling and implemented a system to identify topics in a given document.

In the next chapter, we will learn how to model sequential data using Hidden Markov Models and then use it to analyze stock market data.

11
Probabilistic Reasoning for Sequential Data

In this chapter, we are going to learn how to build sequence learning models. We will learn how to handle time-series data in Pandas. We will understand how to slice time-series data and perform various operations on it. We will discuss how to extract various stats from time-series data on a rolling basis. We will learn about Hidden Markov Models and then implement a system to build those models. We will understand how to use Conditional Random Fields to analyze sequences of alphabets. We will discuss how to analyze stock market data using the techniques learnt so far.

By the end of this chapter, you will learn about:

- Handling time-series data with Pandas
- Slicing time-series data
- Operating on time-series data
- Extracting statistics from time-series data
- Generating data using Hidden Markov Models
- Identifying alphabet sequences with Conditional Random Fields
- Stock market analysis

Understanding sequential data

In the world of machine learning, we encounter many types of data, such as images, text, video, sensor readings, and so on. Different types of data require different types of modeling techniques. Sequential data refers to data where the ordering is important. Time-series data is a particular manifestation of sequential data.

It is basically time-stamped values obtained from any data source such as sensors, microphones, stock markets, and so on. Time-series data has a lot of important characteristics that need to be modeled in order to effectively analyze the data.

The measurements that we encounter in time-series data are taken at regular time intervals and correspond to predetermined parameters. These measurements are arranged on a timeline for storage, and the order of their appearance is very important. We use this order to extract patterns from the data.

In this chapter, we will see how to build models that describe the given time-series data or any sequence in general. These models are used to understand the behavior of the time series variable. We then use these models to predict the future based on past behavior.

Time-series data analysis is used extensively in financial analysis, sensor data analysis, speech recognition, economics, weather forecasting, manufacturing, and many more. We will explore a variety of scenarios where we encounter time-series data and see how we can build a solution. We will be using a library called `Pandas` to handle all the time-series related operations. We will also use a couple of other useful packages like `hmmlearn` and `pystruct` during this chapter. Make sure you install them before you proceed.

You can install them by running the following commands on your Terminal:

```
$ pip3 install pandas
$ pip3 install hmmlearn
$ pip3 install pystruct
$ pip3 install cvxopt
```

If you get an error when installing `cvxopt`, you will find further instructions at `http://cvxopt.org/install`. Now that you have successfully installed the packages, let's go ahead to the next section.

Handling time-series data with Pandas

Let's get started by learning how to handle time-series data in Pandas. In this section, we will convert a sequence of numbers into time series data and visualize it. Pandas provides options to add timestamps, organize data, and then efficiently operate on it.

Create a new Python file and import the following packages:

```
import numpy as np
import matplotlib.pyplot as plt
import pandas as pd
```

Define a function to read the data from the input file. The parameter `index` indicates the column number that contains the relevant data:

```
def read_data(input_file, index):
    # Read the data from the input file
    input_data = np.loadtxt(input_file, delimiter=',')
```

Define a `lambda` function to convert strings to Pandas date format:

```
    # Lambda function to convert strings to Pandas date format
    to_date = lambda x, y: str(int(x)) + '-' + str(int(y))
```

Use this `lambda` function to get the start date from the first line in the input file:

```
    # Extract the start date
    start = to_date(input_data[0, 0], input_data[0, 1])
```

`Pandas` library needs the end date to be exclusive when we perform operations, so we need to increase the date field in the last line by one month:

```
    # Extract the end date
    if input_data[-1, 1] == 12:
        year = input_data[-1, 0] + 1
        month = 1
    else:
        year = input_data[-1, 0]
        month = input_data[-1, 1] + 1

    end = to_date(year, month)
```

Create a list of indices with dates using the start and end dates with a monthly frequency:

```
    # Create a date list with a monthly frequency
    date_indices = pd.date_range(start, end, freq='M')
```

Create pandas data series using the timestamps:

```
    # Add timestamps to the input data to create time-series data
    output = pd.Series(input_data[:, index], index=date_indices)

    return output
```

Define the main function and specify the input file:

```
if __name__=='__main__':
    # Input filename
    input_file = 'data_2D.txt'
```

Specify the columns that contain the data:

```
# Specify the columns that need to be converted
# into time-series data
indices = [2, 3]
```

Iterate through the columns and read the data in each column:

```
# Iterate through the columns and plot the data
for index in indices:
    # Convert the column to timeseries format
    timeseries = read_data(input_file, index)
```

Plot the time-series data:

```
    # Plot the data
    plt.figure()
    timeseries.plot()
    plt.title('Dimension ' + str(index - 1))

plt.show()
```

The full code is given in the file `timeseries.py`. If you run the code, you will see two screenshots.

The following screenshot indicates the data in the first dimension:

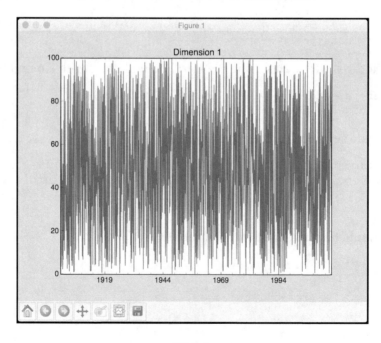

The second screenshot indicates the data in the second dimension:

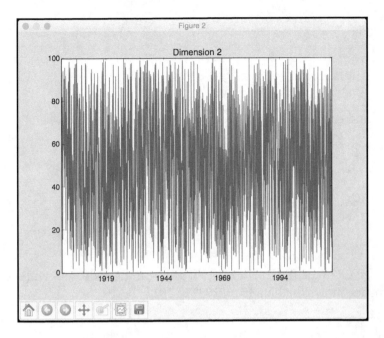

Slicing time-series data

Now that we know how to handle time-series data, let's see how we can slice it. The process of slicing refers to dividing the data into various sub-intervals and extracting relevant information. This is very useful when you are working with time-series datasets. Instead of using indices, we will use timestamp to slice our data.

Create a new Python file and import the following packages:

```
import numpy as np
import matplotlib.pyplot as plt
import pandas as pd

from timeseries import read_data
```

Load the third column (zero-indexed) from the input data file:

```
# Load input data
index = 2
data = read_data('data_2D.txt', index)
```

Define the start and end years, and then plot the data with year-level granularity:

```
# Plot data with year-level granularity
start = '2003'
end = '2011'
plt.figure()
data[start:end].plot()
plt.title('Input data from ' + start + ' to ' + end)
```

Define the start and end months, and then plot the data with month-level granularity:

```
# Plot data with month-level granularity
start = '1998-2'
end = '2006-7'
plt.figure()
data[start:end].plot()
plt.title('Input data from ' + start + ' to ' + end)

plt.show()
```

The full code is given in the file `slicer.py`. If you run the code, you will see two figures. The first screenshot shows the data from *2003* to *2011*:

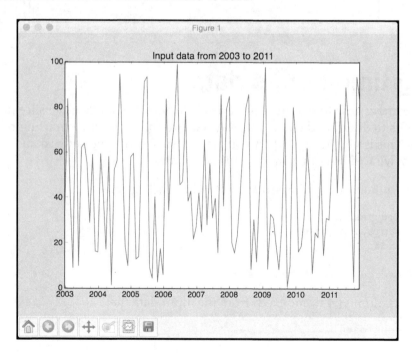

The second screenshot shows the data from *February 1998* to *July 2006*:

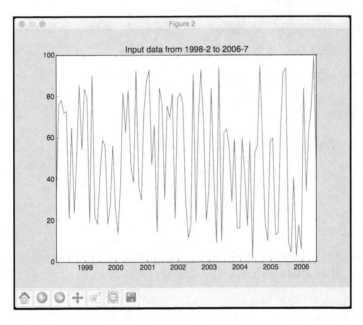

Operating on time-series data

Pandas allows us to operate on time-series data efficiently and perform various operations like filtering and addition. You can simply set some conditions and Pandas will filter the dataset and return the right subset. You can add two time-series variables as well. This allows us to build various applications quickly without having to reinvent the wheel.

Create a new Python file and import the following packages:

```
import numpy as np
import pandas as pd
import matplotlib.pyplot as plt

from timeseries import read_data
```

Define the input filename:

```
# Input filename
input_file = 'data_2D.txt'
```

Load the third and fourth columns into separate variables:

```
# Load data
x1 = read_data(input_file, 2)
x2 = read_data(input_file, 3)
```

Create a Pandas dataframe object by naming the two dimensions:

```
# Create pandas dataframe for slicing
data = pd.DataFrame({'dim1': x1, 'dim2': x2})
```

Plot the data by specifying the start and end years:

```
# Plot data
start = '1968'
end = '1975'
data[start:end].plot()
plt.title('Data overlapped on top of each other')
```

Filter the data using conditions and then display it. In this case, we will take all the datapoints in `dim1` that are less than 45 and all the values in `dim2` that are greater than 30:

```
# Filtering using conditions
# - 'dim1' is smaller than a certain threshold
# - 'dim2' is greater than a certain threshold
data[(data['dim1'] < 45) & (data['dim2'] > 30)].plot()
plt.title('dim1 < 45 and dim2 > 30')
```

We can also add two series in `Pandas`. Let's add `dim1` and `dim2` between the given start and end dates:

```
# Adding two dataframes
plt.figure()
diff = data[start:end]['dim1'] + data[start:end]['dim2']
diff.plot()
plt.title('Summation (dim1 + dim2)')

plt.show()
```

The full code is given in the file `operator.py`. If you run the code, you will see three screenshots. The first screenshot shows the data from *1968* to *1975*:

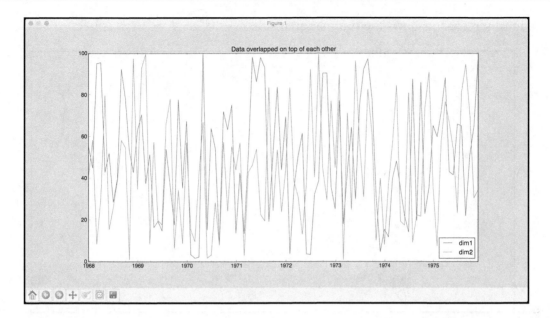

The second screenshot shows the filtered data:

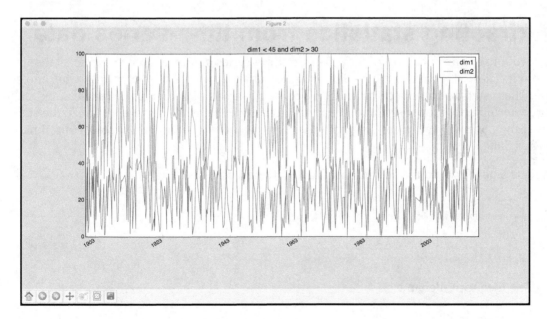

The third screenshot shows the summation result:

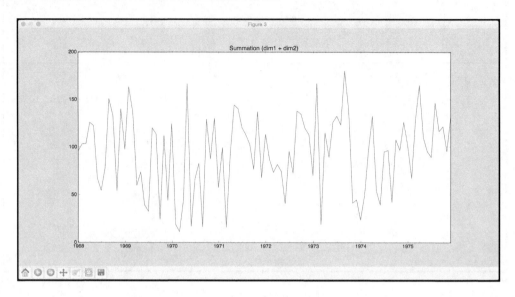

Extracting statistics from time-series data

In order to extract meaningful insights from time-series data, we have to extract statistics from it. These stats can be things like mean, variance, correlation, maximum value, and so on. These stats have to be computed on a rolling basis using a window. We use a predetermined window size and keep computing these stats. When we visualize the stats over time, we will see interesting patterns. Let's see how to extract these stats from time-series data.

Create a new Python file and import the following packages:

```
import numpy as np
import matplotlib.pyplot as plt
import pandas as pd

from timeseries import read_data
```

Define the input filename:

```
# Input filename
input_file = 'data_2D.txt'
```

Load the third and fourth columns into separate variables:

```
# Load input data in time series format
x1 = read_data(input_file, 2)
x2 = read_data(input_file, 3)
```

Create a pandas dataframe by naming the two dimensions:

```
# Create pandas dataframe for slicing
data = pd.DataFrame({'dim1': x1, 'dim2': x2})
```

Extract maximum and minimum values along each dimension:

```
# Extract max and min values
print('\nMaximum values for each dimension:')
print(data.max())
print('\nMinimum values for each dimension:')
print(data.min())
```

Extract the overall mean and the row-wise mean for the first *12* rows:

```
# Extract overall mean and row-wise mean values
print('\nOverall mean:')
print(data.mean())
print('\nRow-wise mean:')
print(data.mean(1)[:12])
```

Plot the rolling mean using a window size of 24:

```
# Plot the rolling mean using a window size of 24
data.rolling(center=False, window=24).mean().plot()
plt.title('Rolling mean')
```

Print the correlation coefficients:

```
# Extract correlation coefficients
print('\nCorrelation coefficients:\n', data.corr())
```

Plot the rolling correlation using a window size of *60*:

```
# Plot rolling correlation using a window size of 60
plt.figure()
plt.title('Rolling correlation')
data['dim1'].rolling(window=60).corr(other=data['dim2']).plot()

plt.show()
```

The full code is given in the file `stats_extractor.py`. If you run the code, you will see two screenshots. The first screenshot shows the rolling mean:

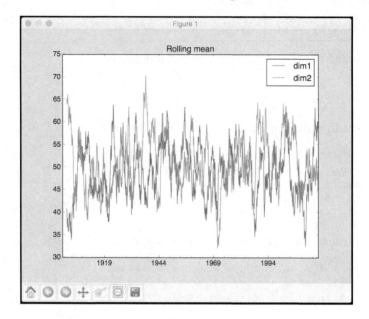

The second screenshot shows the rolling correlation:

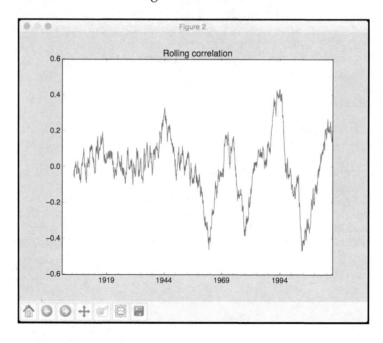

You will see the following on your Terminal:

```
Maximum values for each dimension:
dim1    99.98
dim2    99.97
dtype: float64

Minimum values for each dimension:
dim1    0.18
dim2    0.16
dtype: float64

Overall mean:
dim1    49.030541
dim2    50.983291
dtype: float64
```

If you scroll down, you will see row-wise mean values and the correlation coefficients printed on your Terminal:

```
Row-wise mean:
1900-01-31    85.595
1900-02-28    75.310
1900-03-31    27.700
1900-04-30    44.675
1900-05-31    31.295
1900-06-30    44.160
1900-07-31    67.415
1900-08-31    56.160
1900-09-30    51.495
1900-10-31    61.260
1900-11-30    30.925
1900-12-31    30.785
Freq: M, dtype: float64

Correlation coefficients:
         dim1      dim2
dim1  1.00000   0.00627
dim2  0.00627   1.00000
```

The correlation coefficients in the preceding figures indicate the level of correlation of each dimension with all the other dimensions. A correlation of `1.0` indicates perfect correlation, whereas a correlation of `0.0` indicates that they the variables are not related to each other.

Generating data using Hidden Markov Models

A **Hidden Markov Model** (**HMM**) is a powerful analysis technique for analyzing sequential data. It assumes that the system being modeled is a Markov process with hidden states. This means that the underlying system can be one among a set of possible states. It goes through a sequence of state transitions, thereby producing a sequence of outputs. We can only observe the outputs but not the states. Hence these states are hidden from us. Our goal is to model the data so that we can infer the state transitions of unknown data.

In order to understand HMMs, let's consider the example of a salesman who has to travel between the following three cities for his job — London, Barcelona, and New York. His goal is to minimize the traveling time so that he can be more efficient. Considering his work commitments and schedule, we have a set of probabilities that dictate the chances of going from city X to city Y. In the information given below, $P(X \rightarrow Y)$ indicates the probability of going from city X to city Y:

P(London -> London) = 0.10

P(London -> Barcelona) = 0.70

P(London -> NY) = 0.20

P(Barcelona -> Barcelona) = 0.15

P(Barcelona -> London) = 0.75

P(Barcelona -> NY) = 0.10

P(NY -> NY) = 0.05

P(NY -> London) = 0.60

P(NY -> Barcelona) = 0.35

Let's represent this information with a transition matrix:

London Barcelona NY

London 0.10 0.70 0.20

Barcelona 0.75 0.15 0.10

NY 0.60 0.35 0.05

Now that we have all the information, let's go ahead and set the problem statement. The salesman starts his journey on Tuesday from London and he has to plan something on Friday. But that will depend on where he is. What is the probability that he will be in Barcelona on Friday? This table will help us figure it out.

If we do not have a Markov Chain to model this problem, then we will not know what his travel schedule looks like. Our goal is to say with a good amount of certainty that he will be in a particular city on a given day. If we denote the transition matrix by T and the current day by $X(i)$, then:

$X(i+1) = X(i).T$

In our case, Friday is 3 days away from Tuesday. This means we have to compute $X(i+3)$. The computations will looks like this:

$X(i+1) = X(i).T$

$X(i+2) = X(i+1).T$

$X(i+3) = X(i+2).T$

So in essence:

$X(i+3) = X(i).T^3$

We need to set $X(i)$ as given here:

$X(i) = [0.10\ 0.70\ 0.20]$

The next step is to compute the cube of the matrix. There are many tools available online to perform matrix operations such as `http://matrix.reshish.com/multiplication.php`. If you do all the matrix calculations, then you will see that you will get the following probabilities for Thursday:

$P(London) = 0.31$

$P(Barcelona) = 0.53$

$P(NY) = 0.16$

We can see that there is a higher chance of him being in Barcelona than in any other city. This makes geographical sense as well because Barcelona is closer to London compared to New York. Let's see how to model HMMs in Python.

Create a new Python file and import the following packages:

```
import datetime

import numpy as np
import matplotlib.pyplot as plt
from hmmlearn.hmm import GaussianHMM

from timeseries import read_data
```

Load data from the input file:

```
# Load input data
data = np.loadtxt('data_1D.txt', delimiter=',')
```

Extract the third column for training:

```
# Extract the data column (third column) for training
X = np.column_stack([data[:, 2]])
```

Create a Gaussian HMM with 5 components and diagonal covariance:

```
# Create a Gaussian HMM
num_components = 5
hmm = GaussianHMM(n_components=num_components,
        covariance_type='diag', n_iter=1000)
```

Train the HMM:

```
# Train the HMM
print('\nTraining the Hidden Markov Model...')
hmm.fit(X)
```

Print the mean and variance values for each component of the HMM:

```
# Print HMM stats
print('\nMeans and variances:')
for i in range(hmm.n_components):
    print('\nHidden state', i+1)
    print('Mean =', round(hmm.means_[i][0], 2))
    print('Variance =', round(np.diag(hmm.covars_[i])[0], 2))
```

Generate 1200 samples using the trained HMM model and plot them:

```
# Generate data using the HMM model
num_samples = 1200
generated_data, _ = hmm.sample(num_samples)
plt.plot(np.arange(num_samples), generated_data[:, 0], c='black')
plt.title('Generated data')
```

```
plt.show()
```

The full code is given in the file `hmm.py`. If you run the code, you will see the following screenshot that shows the 1200 generated samples:

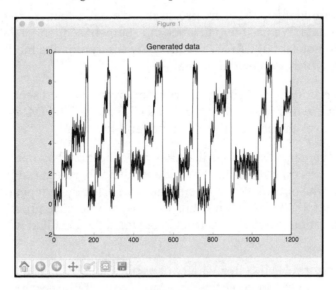

You will see the following printed on your Terminal:

```
Training the Hidden Markov Model...

Means and variances:

Hidden state 1
Mean = 4.6
Variance = 0.25

Hidden state 2
Mean = 6.59
Variance = 0.25

Hidden state 3
Mean = 0.6
Variance = 0.25

Hidden state 4
Mean = 8.6
Variance = 0.26

Hidden state 5
Mean = 2.6
Variance = 0.26
```

Identifying alphabet sequences with Conditional Random Fields

Conditional Random Fields (**CRFs**) are probabilistic models that are frequently used to analyze structured data. We use them to label and segment sequential data in various forms. One thing to note about CRFs is that they are discriminative models. This is in contrast to HMMs, which are generative models.

We can define a conditional probability distribution over a labeled sequence of measurements. We use this framework to build a CRF model. In HMMs, we have to define a joint distribution over the observation sequence and the labels.

One of the main advantages of CRFs is that they are conditional by nature. This is not the case with HMMs. CRFs do not assume any independence between output observations. HMMs assume that the output at any given time is statistically independent of the previous outputs. HMMs need this assumption to ensure that the inference process works in a robust way. But this assumption is not always true! Real world data is filled with temporal dependencies.

CRFs tend to outperform HMMs in a variety of applications such as natural language processing, speech recognition, biotechnology, and so on. In this section, we will discuss how to use CRFs to analyze sequences of alphabets. Create a new python file and import the following packages:

```
import os
import argparse
import string
import pickle

import numpy as np
import matplotlib.pyplot as plt
from pystruct.datasets import load_letters
from pystruct.models import ChainCRF
from pystruct.learners import FrankWolfeSSVM
```

Define a function to parse the input arguments. We can pass the C value as the input parameter. The C parameter controls how much we want to penalize misclassification. A higher value of C would mean that we are imposing a higher penalty for misclassification during training, but we might end up overfitting the model. On the other hand, if we choose a lower value for C, we are allowing the model to generalize well. But this also means that we are imposing a lower penalty for misclassification during training data points.

```
def build_arg_parser():
    parser = argparse.ArgumentParser(description='Trains a Conditional\
            Random Field classifier')
    parser.add_argument("--C", dest="c_val", required=False, type=float,
            default=1.0, help='C value to be used for training')
    return parser
```

Define a class to handle all the functionality of building the CRF model. We will use a chain CRF model with `FrankWolfeSSVM`:

```
# Class to model the CRF
class CRFModel(object):
    def __init__(self, c_val=1.0):
        self.clf = FrankWolfeSSVM(model=ChainCRF(),
                C=c_val, max_iter=50)
```

Define a function to load the training data:

```
    # Load the training data
    def load_data(self):
        alphabets = load_letters()
        X = np.array(alphabets['data'])
        y = np.array(alphabets['labels'])
        folds = alphabets['folds']

        return X, y, folds
```

Define a function to train the CRF model:

```
    # Train the CRF
    def train(self, X_train, y_train):
        self.clf.fit(X_train, y_train)
```

Define a function to evaluate the accuracy of the CRF model:

```
    # Evaluate the accuracy of the CRF
    def evaluate(self, X_test, y_test):
        return self.clf.score(X_test, y_test)
```

Define a function to run the trained CRF model on an unknown datapoint:

```
    # Run the CRF on unknown data
    def classify(self, input_data):
        return self.clf.predict(input_data)[0]
```

Define a function to extract a substring from the alphabets based on a list of indices:

```
# Convert indices to alphabets
def convert_to_letters(indices):
```

```
# Create a numpy array of all alphabets
alphabets = np.array(list(string.ascii_lowercase))
```

Extract the letters:

```
# Extract the letters based on input indices
output = np.take(alphabets, indices)
output = ''.join(output)

return output
```

Define the main function and parse the input arguments:

```
if __name__=='__main__':
    args = build_arg_parser().parse_args()
    c_val = args.c_val
```

Create the CRF model object:

```
# Create the CRF model
crf = CRFModel(c_val)
```

Load the input data and separate it into train and test sets:

```
# Load the train and test data
X, y, folds = crf.load_data()
X_train, X_test = X[folds == 1], X[folds != 1]
y_train, y_test = y[folds == 1], y[folds != 1]
```

Train the CRF model:

```
# Train the CRF model
print('\nTraining the CRF model...')
crf.train(X_train, y_train)
```

Evaluate the accuracy of the CRF model and print it:

```
# Evaluate the accuracy
score = crf.evaluate(X_test, y_test)
print('\nAccuracy score =', str(round(score*100, 2)) + '%')
```

Run it on some test datapoints and print the output:

```
indices = range(3000, len(y_test), 200)
for index in indices:
    print("\nOriginal  =", convert_to_letters(y_test[index]))
    predicted = crf.classify([X_test[index]])
    print("Predicted =", convert_to_letters(predicted))
```

The full code is given in the file `crf.py`. If you run the code, you will see the following output on your Terminal:

```
Training the CRF model...

Accuracy score = 77.93%

Original  = rojections
Predicted = rojectiong

Original  = uff
Predicted = ufr

Original  = kiing
Predicted = kiing

Original  = ecompress
Predicted = ecomertig

Original  = uzz
Predicted = vex

Original  = poiling
Predicted = aciting
```

If you scroll to the end, you will see the following on your Terminal:

```
Original  = abulously
Predicted = abuloualy

Original  = ormalization
Predicted = ormalisation

Original  = ake
Predicted = aka

Original  = afeteria
Predicted = ateteria

Original  = obble
Predicted = obble

Original  = hadow
Predicted = habow

Original  = ndustrialized
Predicted = ndusqrialyled

Original  = ympathetically
Predicted = ympnshetically
```

As we can see, it predicts most of the words correctly.

Stock market analysis

We will analyze stock market data in this section using Hidden Markov Models. This is an example where the data is already organized timestamped. We will use the dataset available in the matplotlib package. The dataset contains the stock values of various companies over the years. Hidden Markov models are generative models that can analyze such time series data and extract the underlying structure. We will use this model to analyze stock price variations and generate the outputs.

Create a new python file and import the following packages:

```
import datetime
import warnings

import numpy as np
import matplotlib.pyplot as plt
from matplotlib.finance import quotes_historical_yahoo_ochl\
        as quotes_yahoo
from hmmlearn.hmm import GaussianHMM
```

Load historical stock market quotes from September 4, 1970 to May 17, 2016. You are free to choose any date range you wish.

```
# Load historical stock quotes from matplotlib package
start = datetime.date(1970, 9, 4)
end = datetime.date(2016, 5, 17)
stock_quotes = quotes_yahoo('INTC', start, end)
```

Extract the closing quote each day and the volume of shares traded that day:

```
# Extract the closing quotes everyday
closing_quotes = np.array([quote[2] for quote in stock_quotes])

# Extract the volume of shares traded everyday
volumes = np.array([quote[5] for quote in stock_quotes])[1:]
```

Take the percentage difference of closing quotes each day:

```
# Take the percentage difference of closing stock prices
diff_percentages = 100.0 * np.diff(closing_quotes) / closing_quotes[:-1]
```

Since the differencing reduces the length of the array by 1, you need to adjust the date array too:

```
# Take the list of dates starting from the second value
dates = np.array([quote[0] for quote in stock_quotes], dtype=np.int)[1:]
```

Stack the two data columns to create the training dataset:

```
# Stack the differences and volume values column-wise for training
training_data = np.column_stack([diff_percentages, volumes])
```

Create and train the Gaussian HMM with 7 components and diagonal covariance:

```
# Create and train Gaussian HMM
hmm = GaussianHMM(n_components=7, covariance_type='diag', n_iter=1000)
with warnings.catch_warnings():
    warnings.simplefilter('ignore')
    hmm.fit(training_data)
```

Use the trained HMM model to generate 300 samples. You can choose to generate any number of samples you want.

```
# Generate data using the HMM model
num_samples = 300
samples, _ = hmm.sample(num_samples)
```

Plot the generated values for difference percentages:

```
# Plot the difference percentages
plt.figure()
plt.title('Difference percentages')
plt.plot(np.arange(num_samples), samples[:, 0], c='black')
```

Plot the generated values for volume of shares traded:

```
# Plot the volume of shares traded
plt.figure()
plt.title('Volume of shares')
plt.plot(np.arange(num_samples), samples[:, 1], c='black')
plt.ylim(ymin=0)

plt.show()
```

Probabilistic Reasoning for Sequential Data

The full code is given in the file `stock_market.py`. If you run the code, you will see the following two screenshots. The first screenshot shows the difference percentages generated by the HMM:

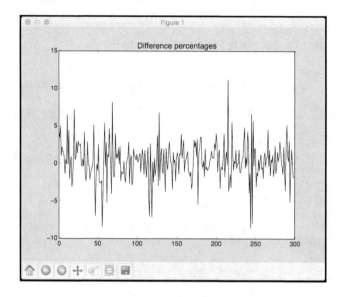

The second screenshot shows the values generated by the HMM for volume of shares traded:

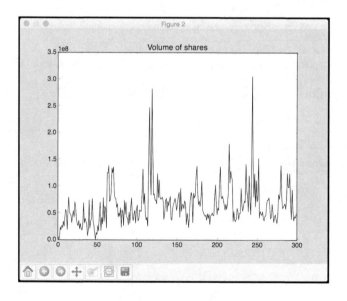

Summary

In this chapter, we learned how to build sequence learning models. We understood how to handle time-series data in Pandas. We discussed how to slice time-series data and perform various operations on it. We learned how to extract various stats from time-series data in a rolling manner. We understood Hidden Markov Models and then implemented a system to build that model.

We discussed how to use Conditional Random Fields to analyze sequences of alphabets. We learned how to analyze stock market data using various techniques. In the next chapter, we will learn about speech recognition and build a system to automatically recognize spoken words.

12
Building A Speech Recognizer

In this chapter, we are going to learn about speech recognition. We will discuss how to work with speech signals and understand how to visualize various audio signals. By utilizing various techniques to process speech signals, we will learn how to build a speech recognition system.

By the end of this chapter, you will know about:

- Working with speech signals
- Visualizing audio signals
- Transforming audio signals to frequency domain
- Generating audio signals
- Synthesizing tones
- Extracting speech features
- Recognizing spoken words

Working with speech signals

Speech recognition is the process of understanding the words that are spoken by humans. The speech signals are captured using a microphone and the system tries to understand the words that are being captured. Speech recognition is used extensively in human computer interaction, smartphones, speech transcription, biometric systems, security, and so on.

It is important to understand the nature of speech signals before we analyze them. These signals happen to be complex mixtures of various signals. There are many different aspects of speech that contribute to its complexity. These things include emotion, accent, language, noise, and so on.

Hence it becomes difficult to robustly define a set of rules to analyze speech signals. But humans are really good at understanding speech even though it has so many variations. We seem to do it with relative ease. If we want our machines to do the same, we need to help them understand speech the same way we do.

Researchers work on various aspects and applications of speech, such as understanding spoken words, identifying who the speaker is, recognizing emotions, identifying accents, and so on. In this chapter, we will focus on understanding spoken words. Speech recognition represents an important step in the field of human computer interaction. If we want to build cognitive robots that can interact with humans, they need to talk to us in natural language. This is the reason that automatic speech recognition has been the center of attention for many researchers in recent years. Let's go ahead and see how to deal with speech signals and build a speech recognizer.

Visualizing audio signals

Let's see how to visualize an audio signal. We will learn how to read an audio signal from a file and work with it. This will help us understand how an audio signal is structured. When audio files are recorded using a microphone, they are sampling the actual audio signals and storing the digitized versions. The real audio signals are continuous valued waves, which means we cannot store them as they are. We need to sample the signal at a certain frequency and convert it into discrete numerical form.

Most commonly, speech signals are sampled at 44,100 Hz. This means that each second of the speech signal is broken down into 44,100 parts and the values at each of these timestamps is stored in an output file. We save the value of the audio signal every 1/44,100 seconds. In this case, we say that the sampling frequency of the audio signal is 44,100 Hz. By choosing a high sampling frequency, it will appear like the audio signal is continuous when humans listen to it. Let's go ahead and visualize an audio signal.

Create a new Python file and import the following packages:

```
import numpy as np
import matplotlib.pyplot as plt
from scipy.io import wavfile
```

Read the input audio file using the `wavfile.read` method. It returns two values – sampling frequency and the audio signal:

```
# Read the audio file
sampling_freq, signal = wavfile.read('random_sound.wav')
```

Print the shape of the signal, datatype, and the duration of the audio signal:

```
# Display the params
print('\nSignal shape:', signal.shape)
print('Datatype:', signal.dtype)
print('Signal duration:', round(signal.shape[0] / float(sampling_freq), 2),
'seconds')
```

Normalize the signal:

```
# Normalize the signal
signal = signal / np.power(2, 15)
```

Extract the first 50 values from the `numpy` array for plotting:

```
# Extract the first 50 values
signal = signal[:50]
```

Construct the time axis in seconds for plotting:

```
# Construct the time axis in milliseconds
time_axis = 1000 * np.arange(0, len(signal), 1) / float(sampling_freq)
```

Plot the audio signal:

```
# Plot the audio signal
plt.plot(time_axis, signal, color='black')
plt.xlabel('Time (milliseconds)')
plt.ylabel('Amplitude')
plt.title('Input audio signal')
plt.show()
```

The full code is given in the file `audio_plotter.py`. If you run the code, you will see the following screenshot:

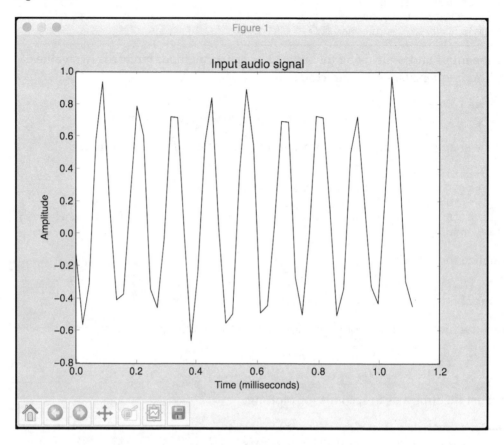

The preceding screenshot shows the first 50 samples of the input audio signal. You will see the following output on your Terminal:

```
Signal shape: (132300,)
Datatype: int16
Signal duration: 3.0 seconds
```

The output printed in the preceding figure shows the information that we extracted from the signal.

Transforming audio signals to the frequency domain

In order to analyze audio signals, we need to understand the underlying frequency components. This gives us insights into how to extract meaningful information from this signal. Audio signals are composed of a mixture of sine waves of varying frequencies, phases, and amplitudes.

If we dissect the frequency components, we can identify a lot of characteristics. Any given audio signal is characterized by its distribution in the frequency spectrum. In order to convert a time domain signal into the frequency domain, we need to use a mathematical tool like Fourier Transform. If you need a quick refresher on **Fourier Transform**, you can check out this link: http://www.thefouriertransform.com. Let's see how to transform an audio signal from the time domain to the frequency domain.

Create a new Python file and import the following packages:

```
import numpy as np
import matplotlib.pyplot as plt
from scipy.io import wavfile
```

Read the input audio file using the `wavefile.read` method. It returns two values – sampling frequency and the audio signal:

```
# Read the audio file
sampling_freq, signal = wavfile.read('spoken_word.wav')
```

Normalize the audio signal:

```
# Normalize the values
signal = signal / np.power(2, 15)
```

Extract the length and half-length of the signal:

```
# Extract the length of the audio signal
len_signal = len(signal)

# Extract the half length
len_half = np.ceil((len_signal + 1) / 2.0).astype(np.int)
```

Apply Fourier transform to the signal:

```
# Apply Fourier transform
freq_signal = np.fft.fft(signal)
```

Normalize the frequency domain signal and take the square:

```
# Normalization
freq_signal = abs(freq_signal[0:len_half]) / len_signal

# Take the square
freq_signal **= 2
```

Adjust the Fourier transformed signal for even and odd cases:

```
# Extract the length of the frequency transformed signal
len_fts = len(freq_signal)

# Adjust the signal for even and odd cases
if len_signal % 2:
    freq_signal[1:len_fts] *= 2
else:
    freq_signal[1:len_fts-1] *= 2
```

Extract the power signal in dB:

```
# Extract the power value in dB
signal_power = 10 * np.log10(freq_signal)
```

Build the X axis, which is frequency measured in kHz in this case:

```
# Build the X axis
x_axis = np.arange(0, len_half, 1) * (sampling_freq / len_signal) / 1000.0
```

Plot the figure:

```
# Plot the figure
plt.figure()
plt.plot(x_axis, signal_power, color='black')
plt.xlabel('Frequency (kHz)')
plt.ylabel('Signal power (dB)')
plt.show()
```

The full code is given in the file `frequency_transformer.py`. If you run the code, you will see the following screenshot:

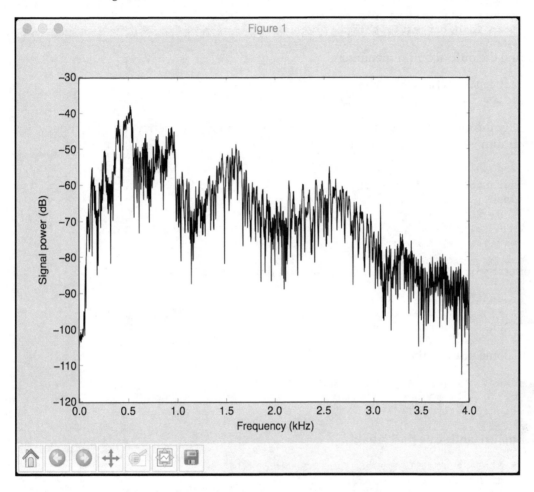

The preceding screenshot shows the power of the signal across the frequency spectrum.

Generating audio signals

Now that we know how audio signals work, let's see how we can generate one such signal. We can use the `NumPy` package to generate various audio signals. Since audio signals are mixtures of **sinusoids,** we can use this to generate an audio signal with some predefined parameters.

Create a new Python file and import the following packages:

```
import numpy as np
import matplotlib.pyplot as plt
from scipy.io.wavfile import write
```

Define the output audio filename:

```
# Output file where the audio will be saved
output_file = 'generated_audio.wav'
```

Specify the audio parameters such as duration, sampling frequency, tone frequency, minimum value, and maximum value:

```
# Specify audio parameters
duration = 4  # in seconds
sampling_freq = 44100  # in Hz
tone_freq = 784
min_val = -4 * np.pi
max_val = 4 * np.pi
```

Generate the audio signal using the defined parameters:

```
# Generate the audio signal
t = np.linspace(min_val, max_val, duration * sampling_freq)
signal = np.sin(2 * np.pi * tone_freq * t)
```

Add some noise to the signal:

```
# Add some noise to the signal
noise = 0.5 * np.random.rand(duration * sampling_freq)
signal += noise
```

Normalize and scale the signal:

```
# Scale it to 16-bit integer values
scaling_factor = np.power(2, 15) - 1
signal_normalized = signal / np.max(np.abs(signal))
signal_scaled = np.int16(signal_normalized * scaling_factor)
```

Save the generated audio signal in the output file:

```
# Save the audio signal in the output file
write(output_file, sampling_freq, signal_scaled)
```

Extract the first 200 values for plotting:

```
# Extract the first 200 values from the audio signal
signal = signal[:200]
```

Construct the time axis in milliseconds:

```
# Construct the time axis in milliseconds
time_axis = 1000 * np.arange(0, len(signal), 1) / float(sampling_freq)
```

Plot the audio signal:

```
# Plot the audio signal
plt.plot(time_axis, signal, color='black')
plt.xlabel('Time (milliseconds)')
plt.ylabel('Amplitude')
plt.title('Generated audio signal')
plt.show()
```

The full code is given in the file `audio_generator.py`. If you run the code, you will see the following screenshot:

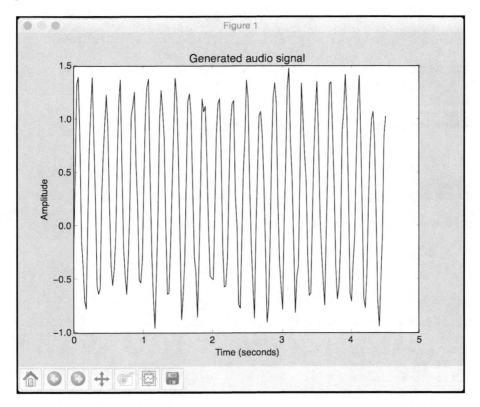

Play the file `generated_audio.wav` using your media player to see what it sounds like. It will be a signal that's a mixture of a *784 Hz* signal and the noise signal.

Synthesizing tones to generate music

The previous section described how to generate a simple monotone, but it's not all that meaningful. It was just a single frequency through the signal. Let's use that principle to synthesize music by stitching different tones together. We will be using standard tones like *A, C, G, F,* and so on to generate music. In order to see the frequency mapping for these standard tones, you can check out this link: `http://www.phy.mtu.edu/~suits/notefreqs.html`. Let's use this information to generate a musical signal.

Create a new Python file and import the following packages:

```
import json

import numpy as np
import matplotlib.pyplot as plt
from scipy.io.wavfile import write
```

Define a function to generate a tone based on the input parameters:

```
# Synthesize the tone based on the input parameters
def tone_synthesizer(freq, duration, amplitude=1.0, sampling_freq=44100):
    # Construct the time axis
    time_axis = np.linspace(0, duration, duration * sampling_freq)
```

Construct the audio signal using the parameters specified and return it:

```
    # Construct the audio signal
    signal = amplitude * np.sin(2 * np.pi * freq * time_axis)

    return signal.astype(np.int16)
```

Define the `main` function. Let's define the output audio filenames:

```
if __name__=='__main__':
    # Names of output files
    file_tone_single = 'generated_tone_single.wav'
    file_tone_sequence = 'generated_tone_sequence.wav'
```

We will be using a tone mapping file that contains the mapping from tone names (such as A, C, G, and so on) to the corresponding frequencies:

```
# Source: http://www.phy.mtu.edu/~suits/notefreqs.html
mapping_file = 'tone_mapping.json'
# Load the tone to frequency map from the mapping file
with open(mapping_file, 'r') as f:
    tone_map = json.loads(f.read())
```

Let's generate the F tone with a duration of 3 seconds:

```
# Set input parameters to generate 'F' tone
tone_name = 'F'
duration = 3      # seconds
amplitude = 12000
sampling_freq = 44100    # Hz
```

Extract the corresponding tone frequency:

```
# Extract the tone frequency
tone_freq = tone_map[tone_name]
```

Generate the tone using the tone synthesizer function that was defined earlier:

```
# Generate the tone using the above parameters
synthesized_tone = tone_synthesizer(tone_freq, duration, amplitude, sampling_freq)
```

Write the generated audio signal to the output file:

```
# Write the audio signal to the output file
write(file_tone_single, sampling_freq, synthesized_tone)
```

Let's generate a tone sequence to make it sound like music. Let's define a tone sequence with corresponding durations in seconds:

```
# Define the tone sequence along with corresponding durations in seconds
tone_sequence = [('G', 0.4), ('D', 0.5), ('F', 0.3), ('C', 0.6), ('A', 0.4)]
```

Construct the audio signal based on the tone sequence:

```
# Construct the audio signal based on the above sequence
signal = np.array([])
for item in tone_sequence:
    # Get the name of the tone
    tone_name = item[0]
```

For each tone, extract the corresponding frequency:

```
# Extract the corresponding frequency of the tone
freq = tone_map[tone_name]
```

Extract the corresponding duration:

```
# Extract the duration
duration = item[1]
```

Synthesize the tone using the tone `synthesizer` function:

```
# Synthesize the tone
synthesized_tone = tone_synthesizer(freq, duration, amplitude, sampling_freq)
```

Append it to the main output signal:

```
# Append the output signal
signal = np.append(signal, synthesized_tone, axis=0)
```

Save the main output signal to the output file:

```
# Save the audio in the output file
write(file_tone_sequence, sampling_freq, signal)
```

The full code is given in the file `synthesizer.py`. If you run the code, it will generate two output files — `generated_tone_single.wav` and `generated_tone_sequence.wav`. Play the audio files using a media player to hear what they sound like.

Extracting speech features

We learnt how to convert a time domain signal into the frequency domain. Frequency domain features are used extensively in all the speech recognition systems. The concept we discussed earlier is an introduction to the idea, but real world frequency domain features are a bit more complex. Once we convert a signal into the frequency domain, we need to ensure that it's usable in the form of a feature vector. This is where the concept of **Mel Frequency Cepstral Coefficients** (**MFCCs**) becomes relevant. MFCC is a tool that's used to extract frequency domain features from a given audio signal.

In order to extract the frequency features from an audio signal, MFCC first extracts the power spectrum. It then uses filter banks and a **discrete cosine transform** (**DCT**) to extract the features. If you are interested in exploring MFCC further, you can check out this link: http://practicalcryptography.com/miscellaneous/machine-learning/guide-mel-frequency-cepstral-coefficients-mfccs.

We will be using a package called `python_speech_features` to extract the MFCC features. The package is available here: http://python-speech-features.readthedocs.org/en/latest. For ease of use, the relevant folder has been included with the code bundle. You will see a folder called `features` in the code bundle that contains the relevant files needed to use this package. Let's see how to extract MFCC features.

Create a new Python file and import the following packages:

```
import numpy as np
import matplotlib.pyplot as plt
from scipy.io import wavfile
from features import mfcc, logfbank
```

Read the input audio file and extract the first 10,000 samples for analysis:

```
# Read the input audio file
sampling_freq, signal = wavfile.read('random_sound.wav')

# Take the first 10,000 samples for analysis
signal = signal[:10000]
```

Extract the MFCC:

```
# Extract the MFCC features
features_mfcc = mfcc(signal, sampling_freq)
```

Print the MFCC parameters:

```
# Print the parameters for MFCC
print('\nMFCC:\nNumber of windows =', features_mfcc.shape[0])
print('Length of each feature =', features_mfcc.shape[1])
```

Plot the MFCC features:

```
# Plot the features
features_mfcc = features_mfcc.T
plt.matshow(features_mfcc)
plt.title('MFCC')
```

Extract the filter bank features:

```
# Extract the Filter Bank features
features_fb = logfbank(signal, sampling_freq)
```

Print the parameters for the filter bank:

```
# Print the parameters for Filter Bank
print('\nFilter bank:\nNumber of windows =', features_fb.shape[0])
print('Length of each feature =', features_fb.shape[1])
```

Plot the features:

```
# Plot the features
features_fb = features_fb.T
plt.matshow(features_fb)
plt.title('Filter bank')

plt.show()
```

The full code is given in the file `feature_extractor.py`. If you run the code, you will see two screenshots. The first screenshot shows the MFCC features:

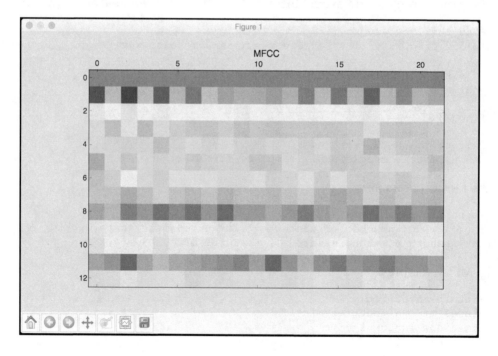

The second screenshot shows the filter bank features:

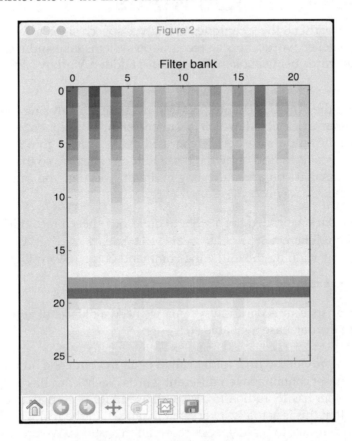

You will see the following printed on your Terminal:

```
MFCC:
Number of windows = 22
Length of each feature = 13

Filter bank:
Number of windows = 22
Length of each feature = 26
```

Recognizing spoken words

Now that we have learnt all the techniques to analyze speech signals, let's go ahead and see how to recognize spoken words. Speech recognition systems take audio signals as input and recognize the words being spoken. We will use Hidden Markov Models (HMMs) for this task.

As we discussed in the previous chapter, HMMs are great at analyzing sequential data. An audio signal is a time series signal, which is a manifestation of sequential data. The assumption is that the outputs are being generated by the system going through a series of hidden states. Our goal is to find out what these hidden states are so that we can identify the words in our signal. If you are interesting in digging deeper, you can check out this link: https://www.robots.ox.ac.uk/~vgg/rg/slides/hmm.pdf.

We will be using a package called `hmmlearn` to build our speech recognition system. You can learn more about it here: http://hmmlearn.readthedocs.org/en/latest. You can install the package by running the following command on your Terminal:

```
$ pip3 install hmmlearn
```

In order to train our speech recognition system, we need a dataset of audio files for each word. We will use the database available at https://code.google.com/archive/p/hmm-speech-recognition/downloads. For ease of use, you have been provided with a folder called `data` in your code bundle that contains all these files. This dataset contains seven different words. Each word has a folder associated with it and each folder has 15 audio files. We will use 14 for training and one for testing in each folder. Note that this is actually a very small dataset. In the real world, you will be using much larger datasets to build speech recognition systems. We are using this dataset to get familiar with speech recognition and see how we can build a system to recognize spoken words.

We will go ahead and build an HMM model for each word. We will store all these models for reference. When we want to recognize the word in an unknown audio file, we will run it through all these models and pick the one with the highest score. Let's see how to build this system.

Create a new Python file and import the following packages:

```
import os
import argparse
import warnings

import numpy as np
from scipy.io import wavfile

from hmmlearn import hmm
from features import mfcc
```

Define a function to parse the input arguments. We need to specify the input folder containing the audio files required to train our speech recognition system:

```
# Define a function to parse the input arguments
def build_arg_parser():
    parser = argparse.ArgumentParser(description='Trains the HMM-based
            speech \ recognition system')
    parser.add_argument("--input-folder", dest="input_folder",
            required=True, help="Input folder containing the audio files
            for training")
    return parser
```

Define a class to train the HMMs:

```
# Define a class to train the HMM
class ModelHMM(object):
    def __init__(self, num_components=4, num_iter=1000):
        self.n_components = num_components
        self.n_iter = num_iter
```

Define the covariance type and the type of HMM:

```
        self.cov_type = 'diag'
        self.model_name = 'GaussianHMM'
```

Initialize the variable in which we will store the models for each word:

```
        self.models = []
```

Define the model using the specified parameters:

```
        self.model = hmm.GaussianHMM(n_components=self.n_components,
                covariance_type=self.cov_type, n_iter=self.n_iter)
```

Define a method to train the model:

```python
# 'training_data' is a 2D numpy array where each row is 13-dimensional
def train(self, training_data):
    np.seterr(all='ignore')
    cur_model = self.model.fit(training_data)
    self.models.append(cur_model)
```

Define a method to compute the score for input data:

```python
# Run the HMM model for inference on input data
def compute_score(self, input_data):
    return self.model.score(input_data)
```

Define a function to build a model for each word in the training dataset:

```python
# Define a function to build a model for each word
def build_models(input_folder):
    # Initialize the variable to store all the models
    speech_models = []
```

Parse the input directory:

```python
# Parse the input directory
for dirname in os.listdir(input_folder):
    # Get the name of the subfolder
    subfolder = os.path.join(input_folder, dirname)

    if not os.path.isdir(subfolder):
        continue
```

Extract the label:

```python
# Extract the label
label = subfolder[subfolder.rfind('/') + 1:]
```

Initialize the variable to store the training data:

```python
# Initialize the variables
X = np.array([])
```

Create a list of files to be used for training:

```python
# Create a list of files to be used for training
# We will leave one file per folder for testing
training_files = [x for x in os.listdir(subfolder) if x.endswith('.wav')][:-1]

# Iterate through the training files and build the models
```

```
    for filename in training_files:
        # Extract the current filepath
        filepath = os.path.join(subfolder, filename)
```

Read the audio signal from the current file:

```
        # Read the audio signal from the input file
        sampling_freq, signal = wavfile.read(filepath)
```

Extract the MFCC features:

```
        # Extract the MFCC features
        with warnings.catch_warnings():
            warnings.simplefilter('ignore')
            features_mfcc = mfcc(signal, sampling_freq)
```

Append the data point to the variable X:

```
        # Append to the variable X
        if len(X) == 0:
            X = features_mfcc
        else:
            X = np.append(X, features_mfcc, axis=0)
```

Initialize the HMM model:

```
        # Create the HMM model
        model = ModelHMM()
```

Train the model using the training data:

```
        # Train the HMM
        model.train(X)
```

Save the model for the current word:

```
        # Save the model for the current word
        speech_models.append((model, label))

        # Reset the variable
        model = None

    return speech_models
```

Define a function to run the tests on the test dataset:

```
# Define a function to run tests on input files
def run_tests(test_files):
    # Classify input data
    for test_file in test_files:
        # Read input file
        sampling_freq, signal = wavfile.read(test_file)
```

Extract the MFCC features:

```
        # Extract MFCC features
        with warnings.catch_warnings():
            warnings.simplefilter('ignore')
            features_mfcc = mfcc(signal, sampling_freq)
```

Define the variables to store the maximum score and the output label:

```
        # Define variables
        max_score = -float('inf')
        output_label = None
```

Iterate through each model to pick the best one:

```
        # Run the current feature vector through all the HMM
        # models and pick the one with the highest score
        for item in speech_models:
            model, label = item
```

Evaluate the score and compare against the maximum score:

```
            score = model.compute_score(features_mfcc)
            if score > max_score:
                max_score = score
                predicted_label = label
```

Print the output:

```
        # Print the predicted output
        start_index = test_file.find('/') + 1
        end_index = test_file.rfind('/')
        original_label = test_file[start_index:end_index]
        print('\nOriginal: ', original_label)
        print('Predicted:', predicted_label)
```

Define the `main` function and get the input folder from the input parameter:

```
if __name__=='__main__':
    args = build_arg_parser().parse_args()
    input_folder = args.input_folder
```

Build an HMM model for each word in the input folder:

```
# Build an HMM model for each word
speech_models = build_models(input_folder)
```

We left one file for testing in each folder. Use that file to see how accurate the model is:

```
# Test files -- the 15th file in each subfolder
test_files = []
for root, dirs, files in os.walk(input_folder):
    for filename in (x for x in files if '15' in x):
        filepath = os.path.join(root, filename)
        test_files.append(filepath)

run_tests(test_files)
```

The full code is given in the file `speech_recognizer.py`. Make sure that the `data` folder is placed in the same folder as the code file. Run the code as given below:

```
$ python3 speech_recognizer.py --input-folder data
```

If you run the code, you will see the following output:

```
Original:  apple
Predicted: apple

Original:  banana
Predicted: banana

Original:  kiwi
Predicted: kiwi

Original:  lime
Predicted: lime

Original:  orange
Predicted: orange

Original:  peach
Predicted: peach

Original:  pineapple
Predicted: pineapple
```

As we can see in the preceding screenshot, our speech recognition system identifies all the words correctly.

Summary

In this chapter, we learnt about speech recognition. We discussed how to work with speech signals and the associated concepts. We learnt how to visualize audio signals. We talked about how to transform time domain audio signals into the frequency domain using Fourier Transforms. We discussed how to generate audio signals using predefined parameters.

We then used this concept to synthesize music by stitching tones together. We talked about MFCCs and how they are used in the real world. We understood how to extract frequency features from speech. We learnt how to use all these techniques to build a speech recognition system. In the next chapter, we will learn about object detection and tracking. We will use those concepts to build an engine that can track objects in a live video.

13
Object Detection and Tracking

In this chapter, we are going to learn about object detection and tracking. We will start by installing OpenCV, a very popular library for computer vision. We will discuss frame differencing to see how we can detect the moving parts in a video. We will learn how to track objects using color spaces. We will understand how to use background subtraction to track objects. We will build an interactive object tracker using the `CAMShift` algorithm. We will learn how to build an optical flow based tracker. We will discuss face detection and associated concepts such as Haar cascades and integral images. We will then use this technique to build an eye detector and tracker.

By the end of this chapter, you will know about:

- Installing OpenCV
- Frame differencing
- Tracking objects using colorspaces
- Object tracking using background subtraction
- Building an interactive object tracker using the CAMShift algorithm
- Optical flow based tracking
- Face detection and tracking
- Using Haar cascades for object detection
- Using integral images for feature extraction
- Eye detection and tracking

Installing OpenCV

We will be using a package called **OpenCV** in this chapter. You can learn more about it here: `http://opencv.org`. Make sure to install it before you proceed. Here are the links to install OpenCV 3 with Python 3 on various operating systems:

- **Windows:**
  `https://solarianprogrammer.com/2016/09/17/install-opencv-3-with-python-3-on-windows`
- **Ubuntu:**
  `http://www.pyimagesearch.com/2015/07/20/install-opencv-3-0-and-python-3-4-on-ubuntu`
- **Mac:**
  `http://www.pyimagesearch.com/2015/06/29/install-opencv-3-0-and-python-3-4-on-osx`

Now that you have installed it, let's go to the next section.

Frame differencing

Frame differencing is one of the simplest techniques that can be used to identify the moving parts in a video. When we are looking at a live video stream, the differences between consecutive frames captured from the stream gives us a lot of information. Let's see how we can take the differences between consecutive frames and display the differences. The code in this section requires an attached camera, so make sure you have a camera on your machine.

Create a new Python file and import the following package:

```
import cv2
```

Define a function to compute the frame differences. Start by computing the difference between the current frame and the next frame:

```
# Compute the frame differences
def frame_diff(prev_frame, cur_frame, next_frame):
    # Difference between the current frame and the next frame
    diff_frames_1 = cv2.absdiff(next_frame, cur_frame)
```

Compute the difference between the current frame and the previous frame:

```
# Difference between the current frame and the previous frame
diff_frames_2 = cv2.absdiff(cur_frame, prev_frame)
```

Compute the bitwise-AND between the two difference frames and return it:

```
return cv2.bitwise_and(diff_frames_1, diff_frames_2)
```

Define a function to grab the current frame from the webcam. Start by reading it from the video capture object:

```
# Define a function to get the current frame from the webcam
def get_frame(cap, scaling_factor):
    # Read the current frame from the video capture object
    _, frame = cap.read()
```

Resize the frame based on the scaling factor and return it:

```
# Resize the image
frame = cv2.resize(frame, None, fx=scaling_factor,
        fy=scaling_factor, interpolation=cv2.INTER_AREA)
```

Convert the image to grayscale and return it:

```
# Convert to grayscale
gray = cv2.cvtColor(frame, cv2.COLOR_RGB2GRAY)

return gray
```

Define the main function and initialize the video capture object:

```
if __name__=='__main__':
    # Define the video capture object
    cap = cv2.VideoCapture(0)
```

Define the scaling factor to resize the images:

```
# Define the scaling factor for the images
scaling_factor = 0.5
```

Object Detection and Tracking

Grab the current frame, the next frame, and the frame after that:

```
# Grab the current frame
prev_frame = get_frame(cap, scaling_factor)
# Grab the next frame
cur_frame = get_frame(cap, scaling_factor)
# Grab the frame after that
next_frame = get_frame(cap, scaling_factor)
```

Iterate indefinitely until the user presses the *Esc* key. Start by computing the frame differences:

```
# Keep reading the frames from the webcam
# until the user hits the 'Esc' key
while True:
    # Display the frame difference
    cv2.imshow('Object Movement', frame_diff(prev_frame,
            cur_frame, next_frame))
```

Update the frame variables:

```
# Update the variables
prev_frame = cur_frame
cur_frame = next_frame
```

Grab the next frame from the webcam:

```
# Grab the next frame
next_frame = get_frame(cap, scaling_factor)
```

Check if the user pressed the *Esc* key. If so, exit the loop:

```
# Check if the user hit the 'Esc' key
key = cv2.waitKey(10)
if key == 27:
    break
```

Once you exit the loop, make sure that all the windows are closed properly:

```
# Close all the windows
cv2.destroyAllWindows()
```

The full code is given in the file `frame_diff.py` provided to you. If you run the code, you will see an output window showing a live output. If you move around, you will see your silhouette as shown here:

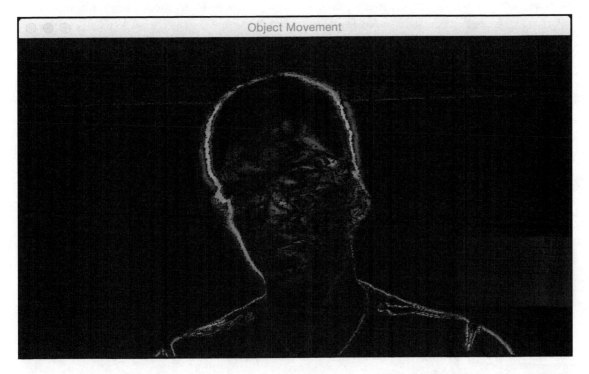

The white lines in the preceding screenshot represent the silhouette.

Tracking objects using colorspaces

The information obtained by frame differencing is useful, but we will not be able to build a robust tracker with it. It is very sensitive to noise and it does not really track an object completely. To build a robust object tracker, we need to know what characteristics of the object can be used to track it accurately. This is where color spaces become relevant.

An image can be represented using various color spaces. The RGB color space is probably the most popular color space, but it does not lend itself nicely to applications like object tracking. So we will be using the HSV color space instead. It is an intuitive color space model that is closer to how humans perceive color. You can learn more about it here: http://infohost.nmt.edu/tcc/help/pubs/colortheory/web/hsv.html. We can convert the captured frame from RGB to HSV colorspace, and then use color thresholding to track any given object. We should note that we need to know the color distribution of the object so that we can select the appropriate ranges for thresholding.

Create a new Python file and import the following packages:

```
import cv2
import numpy as np
```

Define a function to grab the current frame from the webcam. Start by reading it from the video capture object:

```
# Define a function to get the current frame from the webcam
def get_frame(cap, scaling_factor):
    # Read the current frame from the video capture object
    _, frame = cap.read()
```

Resize the frame based on the scaling factor and return it:

```
    # Resize the image
    frame = cv2.resize(frame, None, fx=scaling_factor,
            fy=scaling_factor, interpolation=cv2.INTER_AREA)

    return frame
```

Define the main function. Start by initializing the video capture object:

```
if __name__=='__main__':
    # Define the video capture object
    cap = cv2.VideoCapture(0)
```

Define the scaling factor to be used to resize the captured frames:

```
    # Define the scaling factor for the images
    scaling_factor = 0.5
```

Iterate indefinitely until the user hits the *Esc* key. Grab the current frame to start:

```
# Keep reading the frames from the webcam
# until the user hits the 'Esc' key
while True:
    # Grab the current frame
    frame = get_frame(cap, scaling_factor)
```

Convert the image to HSV color space using the inbuilt function available in OpenCV:

```
# Convert the image to HSV colorspace
hsv = cv2.cvtColor(frame, cv2.COLOR_BGR2HSV)
```

Define the approximate HSV color range for the color of human skin:

```
# Define range of skin color in HSV
lower = np.array([0, 70, 60])
upper = np.array([50, 150, 255])
```

Threshold the HSV image to create the mask:

```
# Threshold the HSV image to get only skin color
mask = cv2.inRange(hsv, lower, upper)
```

Compute bitwise-AND between the mask and the original image:

```
# Bitwise-AND between the mask and original image
img_bitwise_and = cv2.bitwise_and(frame, frame, mask=mask)
```

Run median blurring to smoothen the image:

```
# Run median blurring
img_median_blurred = cv2.medianBlur(img_bitwise_and, 5)
```

Display the input and output frames:

```
# Display the input and output
cv2.imshow('Input', frame)
cv2.imshow('Output', img_median_blurred)
```

Check if the user pressed the *Esc* key. If so, then exit the loop:

```
# Check if the user hit the 'Esc' key
c = cv2.waitKey(5)
if c == 27:
    break
```

Once you exit the loop, make sure that all the windows are properly closed:

```
# Close all the windows
cv2.destroyAllWindows()
```

The full code is given in the file `colorspaces.py` provided to you. If you run the code, you will get two screenshot. The window titled **Input** is the captured frame:

The second window titled **Output** shows the skin mask:

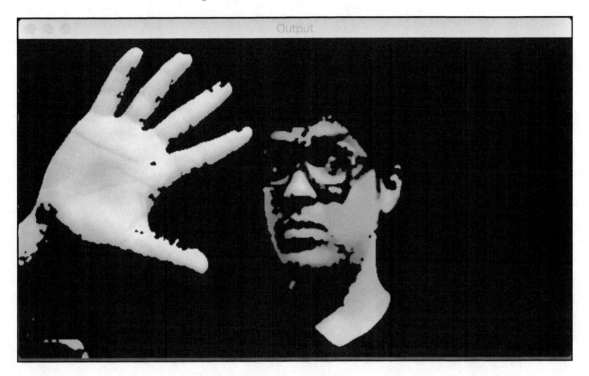

Object tracking using background subtraction

Background subtraction is a technique that models the background in a given video, and then uses that model to detect moving objects. This technique is used a lot in video compression as well as video surveillance. It performs really well where we have to detect moving objects within a static scene. The algorithm basically works by detecting the background, building a model for it, and then subtracting it from the current frame to obtain the foreground. This foreground corresponds to moving objects.

One of the main steps here it to build a model of the background. It is not the same as frame differencing because we are not differencing successive frames. We are actually modeling the background and updating it in real time, which makes it an adaptive algorithm that can adjust to a moving baseline. This is why it performs much better than frame differencing.

Create a new Python file and import the following packages:

```
import cv2
import numpy as np
```

Define a function to grab the current frame:

```
# Define a function to get the current frame from the webcam
def get_frame(cap, scaling_factor):
    # Read the current frame from the video capture object
    _, frame = cap.read()
```

Resize the frame and return it:

```
    # Resize the image
    frame = cv2.resize(frame, None, fx=scaling_factor,
            fy=scaling_factor, interpolation=cv2.INTER_AREA)

    return frame
```

Define the main function and initialize the video capture object:

```
if __name__=='__main__':
    # Define the video capture object
    cap = cv2.VideoCapture(0)
```

Define the background subtractor object:

```
    # Define the background subtractor object
    bg_subtractor = cv2.createBackgroundSubtractorMOG2()
```

Define the history and the learning rate. The comment below is pretty self explanatory as to what "history" is all about:

```
    # Define the number of previous frames to use to learn.
    # This factor controls the learning rate of the algorithm.
    # The learning rate refers to the rate at which your model
    # will learn about the background. Higher value for
    # 'history' indicates a slower learning rate. You can
    # play with this parameter to see how it affects the output.
    history = 100

    # Define the learning rate
    learning_rate = 1.0/history
```

Iterate indefinitely until the user presses the *Esc* key. Start by grabbing the current frame:

```
# Keep reading the frames from the webcam
# until the user hits the 'Esc' key
while True:
    # Grab the current frame
    frame = get_frame(cap, 0.5)
```

Compute the mask using the background subtractor object defined earlier:

```
# Compute the mask
mask = bg_subtractor.apply(frame, learningRate=learning_rate)
```

Convert the mask from grayscale to RGB:

```
# Convert grayscale image to RGB color image
mask = cv2.cvtColor(mask, cv2.COLOR_GRAY2BGR)
```

Display the input and output images:

```
# Display the images
cv2.imshow('Input', frame)
cv2.imshow('Output', mask & frame)
```

Check if the user pressed the *Esc* key. If so, exit the loop:

```
# Check if the user hit the 'Esc' key
c = cv2.waitKey(10)
if c == 27:
    break
```

Once you exit the loop, make sure you release the video capture object and close all the windows properly:

```
# Release the video capture object
cap.release()
# Close all the windows
cv2.destroyAllWindows()
```

The full code is given in the file `background_subtraction.py` provided to you. If you run the code, you will see a window displaying the live output. If you move around, you will partially see yourself as shown here:

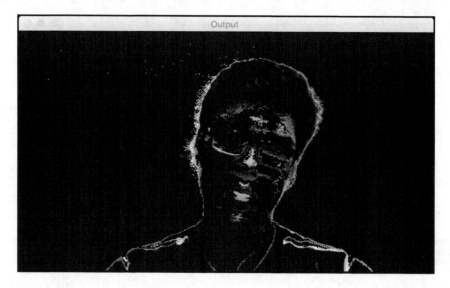

Once you stop moving around, it will start fading because you are now part of the background. The algorithm will consider you a part of the background and start updating the model accordingly:

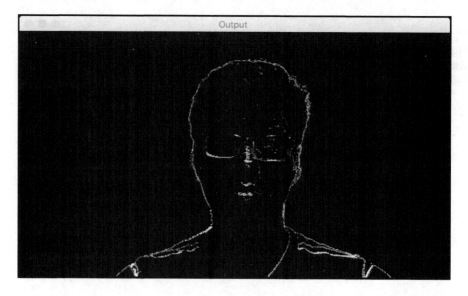

As you remain still, it will continue to fade as shown here:

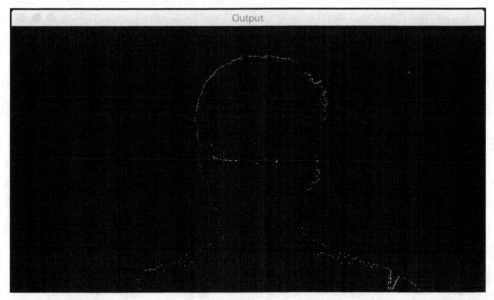

The process of fading indicates that the current scene is becoming part of the background model.

Building an interactive object tracker using the CAMShift algorithm

Color space based tracking allows us to track colored objects, but we have to define the color first. This seems restrictive! Let us see how we can select an object in a live video and then have a tracker that can track it. This is where the CAMShift algorithm, which stands for Continuously Adaptive Mean Shift, becomes relevant. This is basically an adaptive version of the Mean Shift algorithm.

In order to understand **CAMShift,** let's see how Mean Shift works. Consider a region of interest in a given frame. We have selected this region because it contains the object of interest. We want to track this object, so we have drawn a rough boundary around it, which is what "region of interest" refers to. We want our object tracker to track this object as it moves around in the video.

To do this, we select a set of points based on the color histogram of that region and then compute the centroid. If the location of this centroid is at the geometric center of this region, then we know that the object hasn't moved. But if the location of the centroid is not at the geometric center of this region, then we know that the object has moved. This means that we need to move the enclosing boundary as well. The movement of the centroid is directly indicative of the direction of movement of the object. We need to move our bounding box so that the new centroid becomes the geometric center of this bounding box. We keep doing this for every frame, and track the object in real time. Hence, this algorithm is called Mean Shift because the mean (i.e. the centroid) keeps shifting and we track the object using this.

Let us see how this is related to CAMShift. One of the problems with Mean Shift is that the size of the object is not allowed to change over time. Once we draw a bounding box, it will stay constant regardless of how close or far away the object is from the camera. This is why we need to use CAMShift because it can adapt the size of the bounding box to the size of the object. If you want to explore it further, you can check out this link: http://docs.opencv.org/3.1.0/db/df8/tutorial_py_meanshift.html. Let us see how to build a tracker.

Create a new python file and import the following packages:

```
import cv2
import numpy as np
```

Define a class to handle all the functionality related to object tracking:

```
# Define a class to handle object tracking related functionality
class ObjectTracker(object):
    def __init__(self, scaling_factor=0.5):
        # Initialize the video capture object
        self.cap = cv2.VideoCapture(0)
```

Capture the current frame:

```
        # Capture the frame from the webcam
        _, self.frame = self.cap.read()
```

Set the scaling factor:

```
        # Scaling factor for the captured frame
        self.scaling_factor = scaling_factor
```

Resize the frame:

```
# Resize the frame
self.frame = cv2.resize(self.frame, None,
        fx=self.scaling_factor, fy=self.scaling_factor,
        interpolation=cv2.INTER_AREA)
```

Create a window to display the output:

```
# Create a window to display the frame
cv2.namedWindow('Object Tracker')
```

Set the mouse callback function to take input from the mouse:

```
# Set the mouse callback function to track the mouse
cv2.setMouseCallback('Object Tracker', self.mouse_event)
```

Initialize variables to track the rectangular selection:

```
# Initialize variable related to rectangular region selection
self.selection = None
# Initialize variable related to starting position
self.drag_start = None
# Initialize variable related to the state of tracking
self.tracking_state = 0
```

Define a function to track the mouse events:

```
# Define a method to track the mouse events
def mouse_event(self, event, x, y, flags, param):
    # Convert x and y coordinates into 16-bit numpy integers
    x, y = np.int16([x, y])
```

When the left button on the mouse is down, it indicates that the user has started drawing a rectangle:

```
# Check if a mouse button down event has occurred
if event == cv2.EVENT_LBUTTONDOWN:
    self.drag_start = (x, y)
    self.tracking_state = 0
```

If the user is currently dragging the mouse to set the size of the rectangular selection, track the width and height:

```
# Check if the user has started selecting the region
if self.drag_start:
    if flags & cv2.EVENT_FLAG_LBUTTON:
        # Extract the dimensions of the frame
        h, w = self.frame.shape[:2]
```

Set the starting X and Y coordinates of the rectangle:

```
# Get the initial position
xi, yi = self.drag_start
```

Get the maximum and minimum values of the coordinates to make it agnostic to the direction in which you drag the mouse to draw the rectangle:

```
# Get the max and min values
x0, y0 = np.maximum(0, np.minimum([xi, yi], [x, y]))
x1, y1 = np.minimum([w, h], np.maximum([xi, yi], [x, y]))
```

Reset the selection variable:

```
# Reset the selection variable
self.selection = None
```

Finalize the rectangular selection:

```
# Finalize the rectangular selection
if x1-x0 > 0 and y1-y0 > 0:
    self.selection = (x0, y0, x1, y1)
```

If the selection is done, set the flag that indicates that we should start tracking the object within the rectangular region:

```
else:
    # If the selection is done, start tracking
    self.drag_start = None
    if self.selection is not None:
        self.tracking_state = 1
```

Define a method to track the object:

```
# Method to start tracking the object
def start_tracking(self):
    # Iterate until the user presses the Esc key
    while True:
        # Capture the frame from webcam
        _, self.frame = self.cap.read()
```

Resize the frame:

```
# Resize the input frame
self.frame = cv2.resize(self.frame, None,
        fx=self.scaling_factor, fy=self.scaling_factor,
        interpolation=cv2.INTER_AREA)
```

Create a copy of the frame. We will need it later:

```
# Create a copy of the frame
vis = self.frame.copy()
```

Convert the color space of the frame from RGB to HSV:

```
# Convert the frame to HSV colorspace
hsv = cv2.cvtColor(self.frame, cv2.COLOR_BGR2HSV)
```

Create the mask based on predefined thresholds:

```
# Create the mask based on predefined thresholds
mask = cv2.inRange(hsv, np.array((0., 60., 32.)),
        np.array((180., 255., 255.)))
```

Check if the user has selected the region:

```
# Check if the user has selected the region
if self.selection:
    # Extract the coordinates of the selected rectangle
    x0, y0, x1, y1 = self.selection

    # Extract the tracking window
    self.track_window = (x0, y0, x1-x0, y1-y0)
```

Extract the regions of interest from the HSV image as well as the mask. Compute the histogram of the region of interest based on these:

```
    # Extract the regions of interest
    hsv_roi = hsv[y0:y1, x0:x1]
    mask_roi = mask[y0:y1, x0:x1]


    # Compute the histogram of the region of
    # interest in the HSV image using the mask
    hist = cv2.calcHist( [hsv_roi], [0], mask_roi,
            [16], [0, 180] )
```

Normalize the histogram:

```
# Normalize and reshape the histogram
cv2.normalize(hist, hist, 0, 255, cv2.NORM_MINMAX);
self.hist = hist.reshape(-1)
```

Extract the region of interest from the original frame:

```
# Extract the region of interest from the frame
vis_roi = vis[y0:y1, x0:x1]
```

Compute bitwise-NOT of the region of interest. This is for display purposes only:

```
# Compute the image negative (for display only)
cv2.bitwise_not(vis_roi, vis_roi)
vis[mask == 0] = 0
```

Check if the system is in the tracking mode:

```
# Check if the system in the "tracking" mode
if self.tracking_state == 1:
    # Reset the selection variable
    self.selection = None
```

Compute the histogram backprojection:

```
# Compute the histogram back projection
hsv_backproj = cv2.calcBackProject([hsv], [0],
        self.hist, [0, 180], 1)
```

Compute bitwise-AND between the histogram and the mask:

```
# Compute bitwise AND between histogram
# backprojection and the mask
hsv_backproj &= mask
```

Define termination criteria for the tracker:

```
# Define termination criteria for the tracker
term_crit = (cv2.TERM_CRITERIA_EPS |
        cv2.TERM_CRITERIA_COUNT, 10, 1)
```

Apply the `CAMShift` algorithm to the backprojected histogram:

```
# Apply CAMShift on 'hsv_backproj'
track_box, self.track_window = cv2.CamShift(hsv_backproj,
        self.track_window, term_crit)
```

Draw an ellipse around the object and display it:

```
# Draw an ellipse around the object
cv2.ellipse(vis, track_box, (0, 255, 0), 2)
# Show the output live video
cv2.imshow('Object Tracker', vis)
```

If the user presses *Esc*, then exit the loop:

```
# Stop if the user hits the 'Esc' key
c = cv2.waitKey(5)
if c == 27:
    break
```

Once you exit the loop, make sure that all the windows are closed properly:

```
# Close all the windows
cv2.destroyAllWindows()
```

Define the main function and start tracking:

```
if __name__ == '__main__':
    # Start the tracker
    ObjectTracker().start_tracking()
```

The full code is given in the file `camshift.py` provided to you. If you run the code, you will see a window showing the live video from the webcam.

Take an object, hold it in your hand, and then draw a rectangle around it. Once you draw the rectangle, make sure to move the mouse pointer away from the final position. The image will look something like this:

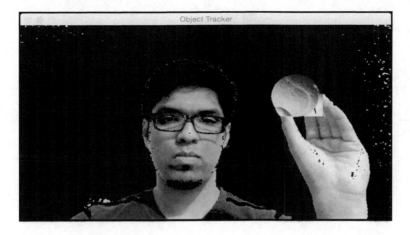

Once the selection is done, move the mouse pointer to a different position to lock the rectangle. This event will start the tracking process as seen in the following image:

Let's move the object around to see if it's still being tracked:

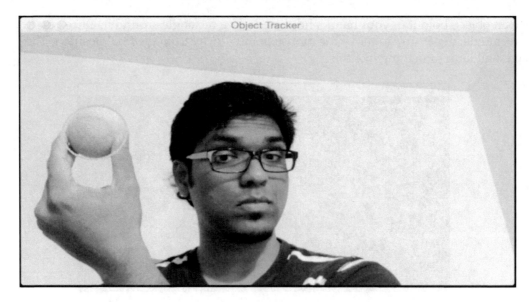

Looks like it's working well. You can move the object around to see how it's getting tracked in real time.

Optical flow based tracking

Optical flow is a very popular technique used in computer vision. It uses image feature points to track an object. Individual feature points are tracked across successive frames in the live video. When we detect a set of feature points in a given frame, we compute the displacement vectors to keep track of it. We show the motion of these feature points between successive frames. These vectors are known as motion vectors. There are many different ways to perform optical flow, but the **Lucas-Kanade** method is perhaps the most popular. Here is the original paper that describes this technique:
http://cseweb.ucsd.edu/classes/sp02/cse252/lucaskanade81.pdf.

The first step is to extract the feature points from the current frame. For each feature point that is extracted, a 3×3 patch (of pixels) is created with the feature point at the center. We are assuming that all the points in each patch have a similar motion. The size of this window can be adjusted depending on the situation.

For each patch, we look for a match in its neighborhood in the previous frame. We pick the best match based on an error metric. The search area is bigger than 3×3 because we look for a bunch of different 3×3 patches to get the one that is closest to the current patch. Once we get that, the path from the center point of the current patch and the matched patch in the previous frame will become the motion vector. We similarly compute the motion vectors for all the other patches.

Create a new python file and import the following packages:

```
import cv2
import numpy as np
```

Define a function to start tracking using optical flow. Start by initializing the video capture object and the scaling factor:

```
# Define a function to track the object
def start_tracking():
    # Initialize the video capture object
    cap = cv2.VideoCapture(0)

    # Define the scaling factor for the frames
    scaling_factor = 0.5
```

Define the number of frames to track and the number of frames to skip:

```
# Number of frames to track
num_frames_to_track = 5

# Skipping factor
num_frames_jump = 2
```

Initialize variables related to tracking paths and frame index:

```
# Initialize variables
tracking_paths = []
frame_index = 0
```

Define the tracking parameters like the window size, maximum level, and the termination criteria:

```
# Define tracking parameters
tracking_params = dict(winSize  = (11, 11), maxLevel = 2,
            criteria = (cv2.TERM_CRITERIA_EPS | cv2.TERM_CRITERIA_COUNT,
                10, 0.03))
```

Iterate indefinitely until the user presses the *Esc* key. Start by capturing the current frame and resizing it:

```
# Iterate until the user hits the 'Esc' key
while True:
    # Capture the current frame
    _, frame = cap.read()

    # Resize the frame
    frame = cv2.resize(frame, None, fx=scaling_factor,
            fy=scaling_factor, interpolation=cv2.INTER_AREA)
```

Convert the frame from RGB to grayscale:

```
    # Convert to grayscale
    frame_gray = cv2.cvtColor(frame, cv2.COLOR_BGR2GRAY)
```

Create a copy of the frame:

```
    # Create a copy of the frame
    output_img = frame.copy()
```

Check if the length of tracking paths is greater than zero:

```
if len(tracking_paths) > 0:
    # Get images
    prev_img, current_img = prev_gray, frame_gray
```

Organize the feature points:

```
    # Organize the feature points
    feature_points_0 = np.float32([tp[-1] for tp in \
            tracking_paths]).reshape(-1, 1, 2)
```

Compute the optical flow based on the previous and current images by using the feature points and the tracking parameters:

```
    # Compute optical flow
    feature_points_1, _, _ = cv2.calcOpticalFlowPyrLK(
            prev_img, current_img, feature_points_0,
            None, **tracking_params)
    # Compute reverse optical flow
    feature_points_0_rev, _, _ = cv2.calcOpticalFlowPyrLK(
            current_img, prev_img, feature_points_1,
            None, **tracking_params)

    # Compute the difference between forward and
    # reverse optical flow
    diff_feature_points = abs(feature_points_0 - \
            feature_points_0_rev).reshape(-1, 2).max(-1)
```

Extract the good feature points:

```
    # Extract the good points
    good_points = diff_feature_points < 1
```

Initialize the variable for the new tracking paths:

```
    # Initialize variable
    new_tracking_paths = []
```

Iterate through all the good feature points and draw circles around them:

```
# Iterate through all the good feature points
for tp, (x, y), good_points_flag in zip(tracking_paths,
            feature_points_1.reshape(-1, 2), good_points):
    # If the flag is not true, then continue
    if not good_points_flag:
        continue
```

Append the X and Y coordinates and don't exceed the number of frames we are supposed to track:

```
    # Append the X and Y coordinates and check if
    # its length greater than the threshold
    tp.append((x, y))
    if len(tp) > num_frames_to_track:
        del tp[0]

    new_tracking_paths.append(tp)
```

Draw a circle around the point. Update the tracking paths and draw lines using the new tracking paths to show movement:

```
    # Draw a circle around the feature points
    cv2.circle(output_img, (x, y), 3, (0, 255, 0), -1)

# Update the tracking paths
tracking_paths = new_tracking_paths

# Draw lines
cv2.polylines(output_img, [np.int32(tp) for tp in \
        tracking_paths], False, (0, 150, 0))
```

Go into this `if` condition after skipping the number of frames specified earlier:

```
# Go into this 'if' condition after skipping the
# right number of frames
if not frame_index % num_frames_jump:
    # Create a mask and draw the circles
    mask = np.zeros_like(frame_gray)
    mask[:] = 255
    for x, y in [np.int32(tp[-1]) for tp in tracking_paths]:
        cv2.circle(mask, (x, y), 6, 0, -1)
```

Compute the good features to track using the inbuilt function along with parameters like mask, maximum corners, quality level, minimum distance, and the block size:

```
# Compute good features to track
feature_points = cv2.goodFeaturesToTrack(frame_gray,
        mask = mask, maxCorners = 500, qualityLevel = 0.3,
        minDistance = 7, blockSize = 7)
```

If the feature points exist, append them to the tracking paths:

```
# Check if feature points exist. If so, append them
# to the tracking paths
if feature_points is not None:
    for x, y in np.float32(feature_points).reshape(-1, 2):
        tracking_paths.append([(x, y)])
```

Update the variables related to frame index and the previous grayscale image:

```
# Update variables
frame_index += 1
prev_gray = frame_gray
```

Display the output:

```
# Display output
cv2.imshow('Optical Flow', output_img)
```

Check if the user pressed the *Esc* key. If so, exit the loop:

```
# Check if the user hit the 'Esc' key
c = cv2.waitKey(1)
if c == 27:
    break
```

Define the main function and start tracking. Once you stop the tracker, make sure that all the windows are closed properly:

```
if __name__ == '__main__':
    # Start the tracker
    start_tracking()

    # Close all the windows
    cv2.destroyAllWindows()
```

The full code is given in the file `optical_flow.py` provided to you. If you run the code, you will see a window showing the live video. You will see feature points as shown in the following screensot:

If you move around, you will see lines showing the movement of those feature points:

If you then move in the opposite direction, the lines will also change their direction accordingly:

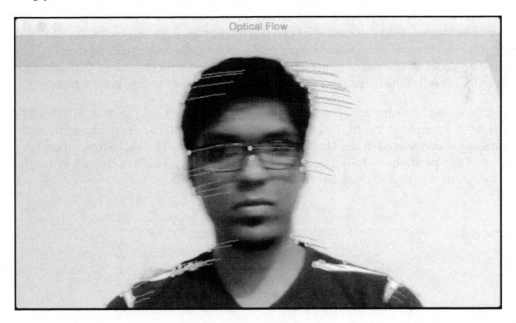

Face detection and tracking

Face detection refers to detecting the location of a face in a given image. This is often confused with face recognition, which is the process of identifying who the person is. A typical biometric system utilizes both face detection and face recognition to perform the task. It uses face detection to locate the face and then uses face recognition to identify the person. In this section, we will see how to automatically detect the location of a face in a live video and track it.

Using Haar cascades for object detection

We will be using Haar cascades to detect faces in the video. Haar cascades, in this case, refer to cascade classifiers based on Haar features. *Paul Viola* and *Michael Jones* first came up with this object detection method in their landmark research paper in 2001. You can check it out here: `https://www.cs.cmu.edu/~efros/courses/LBMV07/Papers/viola-cvpr-01.pdf`. In their paper, they describe an effective machine learning technique that can be used to detect any object.

They use a boosted cascade of simple classifiers. This cascade is used to build an overall classifier that performs with high accuracy. The reason this is relevant is because it helps us circumvent the process of building a single-step classifier that performs with high accuracy. Building one such robust single-step classifier is a computationally intensive process.

Consider an example where we have to detect an object like, say, a tennis ball. In order to build a detector, we need a system that can learn what a tennis ball looks like. It should be able to infer whether or not a given image contains a tennis ball. We need to train this system using a lot of images of tennis balls. We also need a lot of images that don't contain tennis balls as well. This helps the system learn how to differentiate between objects.

If we build an accurate model, it will be complex. Hence we won't be able to run it in real time. If it's too simple, it might not be accurate. This trade off between speed and accuracy is frequently encountered in the world of machine learning. The Viola-Jones method overcomes this problem by building a set of simple classifiers. These classifiers are then cascaded to form a unified classifier that's robust and accurate.

Let's see how to use this to do face detection. In order to build a machine learning system to detect faces, we first need to build a feature extractor. The machine learning algorithms will use these features to understand what a face looks like. This is where Haar features become relevant. They are just simple summations and differences of patches across the image. Haar features are really easy to compute. In order to make it robust to scale, we do this at multiple image sizes. If you want to learn more about this in a tutorial format, you can check out this link: `http://www.cs.ubc.ca/~lowe/425/slides/13-ViolaJones.pdf`.

Once the features are extracted, we pass them through our boosted cascade of simple classifiers. We check various rectangular sub-regions in the image and keep discarding the ones that don't contain faces. This helps us arrive at the final answer quickly. In order to compute these features quickly, they used a concept known as integral images.

Using integral images for feature extraction

In order to compute Haar features, we have to compute the summations and differences of many sub-regions in the image. We need to compute these summations and differences at multiple scales, which makes it a very computationally intensive process. In order to build a real time system, we use integral images. Consider the following figure:

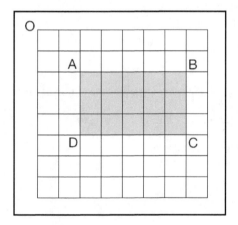

If we want to compute the sum of the rectangle ABCD in this image, we don't need to go through each pixel in that rectangular area. Let's say OP indicates the area of the rectangle formed by the top left corner O and the point P on the diagonally opposite corners of the rectangle. To calculate the area of the rectangle ABCD, we can use the following formula:

Area of the rectangle ABCD = OC − (OB + OD − OA)

What's so special about this formula? If you notice, we didn't have to iterate through anything or recalculate any rectangle areas. All the values on the right hand side of the equation are already available because they were computed during earlier cycles. We directly used them to compute the area of this rectangle. Let's see how to build a face detector.

Create a new python file and import the following packages:

```
import cv2
import numpy as np
```

Load the Haar cascade file corresponding to face detection:

```
# Load the Haar cascade file
face_cascade = cv2.CascadeClassifier(
        'haar_cascade_files/haarcascade_frontalface_default.xml')

# Check if the cascade file has been loaded correctly
if face_cascade.empty():
    raise IOError('Unable to load the face cascade classifier xml file')
```

Initialize the video capture object and define the scaling factor:

```
# Initialize the video capture object
cap = cv2.VideoCapture(0)

# Define the scaling factor
scaling_factor = 0.5
```

Iterate indefinitely until the user presses the *Esc* key. Capture the current frame:

```
# Iterate until the user hits the 'Esc' key
while True:
    # Capture the current frame
    _, frame = cap.read()
```

Resize the frame:

```
    # Resize the frame
    frame = cv2.resize(frame, None,
            fx=scaling_factor, fy=scaling_factor,
            interpolation=cv2.INTER_AREA)
```

Convert the image to grayscale:

```
# Convert to grayscale
gray = cv2.cvtColor(frame, cv2.COLOR_BGR2GRAY)
```

Run the face detector on the grayscale image:

```
# Run the face detector on the grayscale image
face_rects = face_cascade.detectMultiScale(gray, 1.3, 5)
```

Iterate through the detected faces and draw rectangles around them:

```
# Draw a rectangle around the face
for (x,y,w,h) in face_rects:
    cv2.rectangle(frame, (x,y), (x+w,y+h), (0,255,0), 3)
```

Display the output:

```
# Display the output
cv2.imshow('Face Detector', frame)
```

Check if the user pressed the *Esc* key. If so, exit the loop:

```
# Check if the user hit the 'Esc' key
c = cv2.waitKey(1)
if c == 27:
    break
```

Once you exit the loop, make sure to release the video capture object and close all the windows properly:

```
# Release the video capture object
cap.release()

# Close all the windows
cv2.destroyAllWindows()
```

The full code is given in the file `face_detector.py` provided to you. If you run the code, you will see something like this:

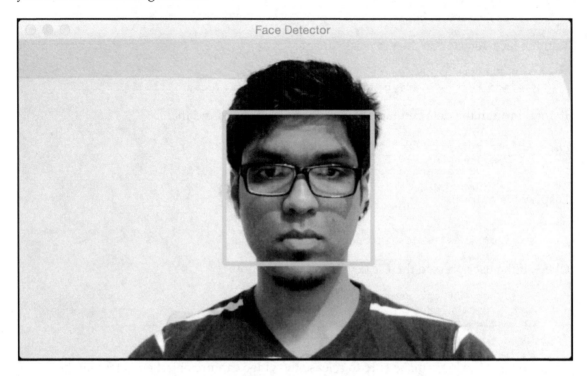

Eye detection and tracking

Eye detection works very similarly to face detection. Instead of using a face cascade file, we will use an eye cascade file. Create a new python file and import the following packages:

```
import cv2
import numpy as np
```

Load the Haar cascade files corresponding to face and eye detection:

```
# Load the Haar cascade files for face and eye
face_cascade = 
cv2.CascadeClassifier('haar_cascade_files/haarcascade_frontalface_default.xml')
eye_cascade = 
cv2.CascadeClassifier('haar_cascade_files/haarcascade_eye.xml')
```

```
# Check if the face cascade file has been loaded correctly
if face_cascade.empty():
    raise IOError('Unable to load the face cascade classifier xml file')

# Check if the eye cascade file has been loaded correctly
if eye_cascade.empty():
    raise IOError('Unable to load the eye cascade classifier xml file')
```

Initialize the video capture object and define the scaling factor:

```
# Initialize the video capture object
cap = cv2.VideoCapture(0)
# Define the scaling factor
ds_factor = 0.5
```

Iterate indefinitely until the user presses the *Esc* key:

```
# Iterate until the user hits the 'Esc' key
while True:
    # Capture the current frame
    _, frame = cap.read()
```

Resize the frame:

```
    # Resize the frame
    frame = cv2.resize(frame, None, fx=ds_factor, fy=ds_factor,
interpolation=cv2.INTER_AREA)
```

Convert the frame from RGB to grayscale:

```
    # Convert to grayscale
    gray = cv2.cvtColor(frame, cv2.COLOR_BGR2GRAY)
```

Run the face detector:

```
    # Run the face detector on the grayscale image
    faces = face_cascade.detectMultiScale(gray, 1.3, 5)
```

For each face detected, run the eye detector within that region:

```
    # For each face that's detected, run the eye detector
    for (x,y,w,h) in faces:
        # Extract the grayscale face ROI
        roi_gray = gray[y:y+h, x:x+w]
```

Object Detection and Tracking

Extract the region of interest and run the eye detector:

```
# Extract the color face ROI
roi_color = frame[y:y+h, x:x+w]

# Run the eye detector on the grayscale ROI
eyes = eye_cascade.detectMultiScale(roi_gray)
```

Draw circles around the eyes and display the output:

```
# Draw circles around the eyes
for (x_eye,y_eye,w_eye,h_eye) in eyes:
    center = (int(x_eye + 0.5*w_eye), int(y_eye + 0.5*h_eye))
    radius = int(0.3 * (w_eye + h_eye))
    color = (0, 255, 0)
    thickness = 3
    cv2.circle(roi_color, center, radius, color, thickness)
# Display the output
cv2.imshow('Eye Detector', frame)
```

If the user presses the *Esc* key, exit the loop:

```
# Check if the user hit the 'Esc' key
c = cv2.waitKey(1)
if c == 27:
    break
```

Once you exit the loop, make sure to release the video capture object and close all the windows:

```
# Release the video capture object
cap.release()

# Close all the windows
cv2.destroyAllWindows()
```

The full code is given in the file `eye_detector.py` provided to you. If you run the code, you will see something like this:

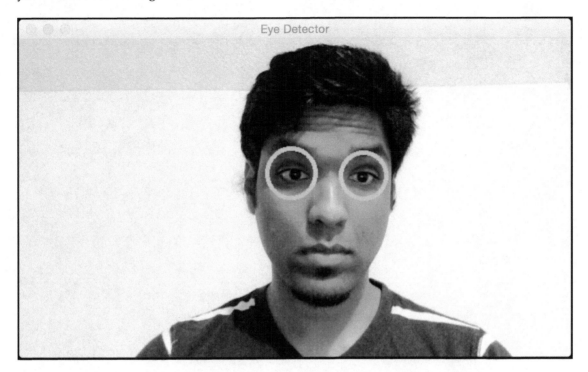

Summary

In this chapter, we learnt about object detection and tracking. We understood how to install OpenCV with Python support on various operating systems. We learnt about frame differencing and used it to detect the moving parts in a video. We discussed how to track human skin using color spaces. We talked about background subtraction and how it can be used to track objects in static scenes. We built an interactive object tracker using the CAMShift algorithm.

We learnt how to build an optical flow based tracker. We discussed face detection techniques and understood the concepts of Haar cascades and integral images. We used this technique to build an eye detector and tracker. In the next chapter, we will discuss artificial neural networks and use those techniques to build an optical character recognition engine.

14
Artificial Neural Networks

In this chapter, we are going to learn about artificial neural networks. We will start with an introduction to artificial neural networks and the installation of the relevant library. We will discuss perceptrons and how to build a classifier based on them. We will learn about single layer neural networks and multilayer neural networks. We will see how to use neural networks to build a vector quantizer. We will analyze sequential data using recurrent neural networks. We will then use artificial neural networks to build an optical character recognition engine.

By the end of this chapter, you will know about:

- Introduction to artificial neural networks
- Building a Perceptron based classifier
- Constructing a single layer neural network
- Constructing a multilayer neural network
- Building a vector quantizer
- Analyzing sequential data using recurrent neural networks
- Visualizing characters in an Optical Character Recognition (OCR) database
- Building an Optical Character Recognition (OCR) engine

Introduction to artificial neural networks

One of the fundamental premises of Artificial Intelligence is to build machines that can perform tasks that require human intelligence. The human brain is amazing at learning new things. Why not use the model of the human brain to build a machine? An artificial neural network is a model designed to simulate the learning process of the human brain.

Artificial neural networks are designed such that they can identify the underlying patterns in data and learn from them. They can be used for various tasks such as classification, regression, segmentation, and so on. We need to convert any given data into numerical form before feeding it into the neural network. For example, we deal with many different types of data including visual, textual, time-series, and so on. We need to figure out how to represent problems in a way that can be understood by artificial neural networks.

Building a neural network

The human learning process is hierarchical. We have various stages in our brain's neural network and each stage corresponds to a different granularity. Some stages learn simple things and some stages learn more complex things. Let's consider an example of visually recognizing an object. When we look at a box, the first stage identifies simple things like corners and edges. The next stage identifies the generic shape and the stage after that identifies what kind of object it is. This process differs for different tasks, but you get the idea! By building this hierarchy, our human brain quickly separates the concepts and identifies the given object.

To simulate the learning process of the human brain, an artificial neural network is built using layers of neurons. These neurons are inspired by the biological neurons we discussed in the previous paragraph. Each layer in an artificial neural network is a set of independent neurons. Each neuron in a layer is connected to neurons in the adjacent layer.

Training a neural network

If we are dealing with N-dimensional input data, then the input layer will consist of N neurons. If we have M distinct classes in our training data, then the output layer will consist of M neurons. The layers between the input and output layers are called hidden layers. A simple neural network will consist of a couple of layers and a deep neural network will consist of many layers.

Consider the case where we want to use a neural network to classify the given data. The first step is to collect the appropriate training data and label it. Each neuron acts as a simple function and the neural network trains itself until the error goes below a certain value. The error is basically the difference between the predicted output and the actual output. Based on how big the error is, the neural network adjusts itself and retrains until it gets closer to the solution. You can learn more about neural networks
at `http://pages.cs.wisc.edu/~bolo/shipyard/neural/local.html`.

We will be using a library called NeuroLab in this chapter. You can find more about it at https://pythonhosted.org/neurolab. You can install it by running the following command on your Terminal:

```
$ pip3 install neurolab
```

Once you have installed it, you can proceed to the next section.

Building a Perceptron based classifier

A **Perceptron** is the building block of an artificial neural network. It is a single neuron that takes inputs, performs computation on them, and then produces an output. It uses a simple linear function to make the decision. Let's say we are dealing with an N-dimension input data point. A Perceptron computes the weighted summation of those N numbers and it then adds a constant to produce the output. The constant is called the bias of the neuron. It is remarkable to note that these simple Perceptrons are used to design very complex deep neural networks. Let's see how to build a Perceptron based classifier using NeuroLab.

Create a new Python file and import the following packages:

```
import numpy as np
import matplotlib.pyplot as plt
import neurolab as nl
```

Load the input data from the text file `data_perceptron.txt` provided to you. Each line contains space separated numbers where the first two numbers are the features and the last number is the label:

```
# Load input data
text = np.loadtxt('data_perceptron.txt')
```

Separate the text into datapoints and labels:

```
# Separate datapoints and labels
data = text[:, :2]
labels = text[:, 2].reshape((text.shape[0], 1))
```

Plot the datapoints:

```
# Plot input data
plt.figure()
plt.scatter(data[:,0], data[:,1])
plt.xlabel('Dimension 1')
plt.ylabel('Dimension 2')
plt.title('Input data')
```

Define the maximum and minimum values that each dimension can take:

```
# Define minimum and maximum values for each dimension
dim1_min, dim1_max, dim2_min, dim2_max = 0, 1, 0, 1
```

Since the data is separated into two classes, we just need one bit to represent the output. So the output layer will contain a single neuron.

```
# Number of neurons in the output layer
num_output = labels.shape[1]
```

We have a dataset where the datapoints are two-dimensional. Let's define a Perceptron with two input neurons, where we assign one neuron for each dimension.

```
# Define a perceptron with 2 input neurons (because we
# have 2 dimensions in the input data)
dim1 = [dim1_min, dim1_max]
dim2 = [dim2_min, dim2_max]
perceptron = nl.net.newp([dim1, dim2], num_output)
```

Train the perceptron with the training data:

```
# Train the perceptron using the data
error_progress = perceptron.train(data, labels, epochs=100, show=20,
lr=0.03)
```

Plot the training progress using the error metric:

```
# Plot the training progress
plt.figure()
plt.plot(error_progress)
plt.xlabel('Number of epochs')
plt.ylabel('Training error')
plt.title('Training error progress')
plt.grid()

plt.show()
```

Chapter 14

The full code is given in the file perceptron_classifier.py. If you run the code, you will get two output screenshot. The first screenshot indicates the input data points:

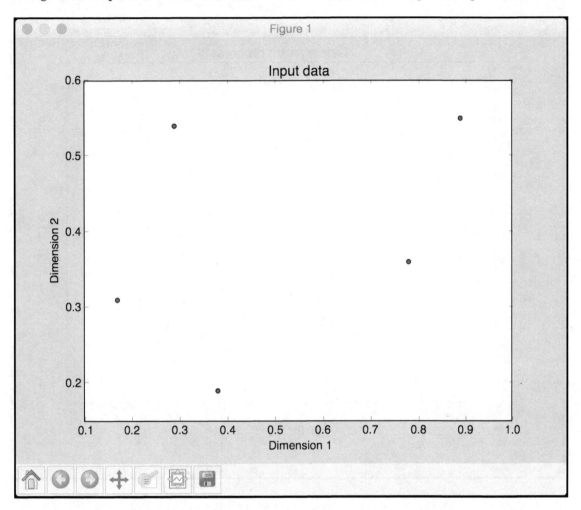

[367]

The second screenshot represents the training progress using the error metric:

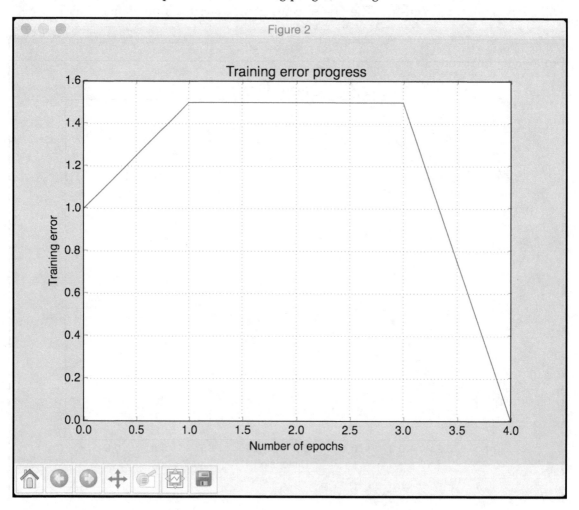

We can observe from the preceding screenshot, the error goes down to **0** at the end of the fourth epoch.

Constructing a single layer neural network

A perceptron is a good start, but it cannot do much. The next step is to have a set of neurons act as a unit to see what we can achieve. Let's create a single neural network that consists of independent neurons acting on input data to produce the output.

Create a new Python file and import the following packages:

```
import numpy as np
import matplotlib.pyplot as plt
import neurolab as nl
```

We will use the input data from the file `data_simple_nn.txt` provided to you. Each line in this file contains four numbers. The first two numbers form the datapoint and the last two numbers are the labels. Why do we need to assign two numbers for labels? Because we have four distinct classes in our dataset, so we need two bits to represent them. Let us go ahead and load the data:

```
# Load input data
text = np.loadtxt('data_simple_nn.txt')
```

Separate the data into datapoints and labels:

```
# Separate it into datapoints and labels
data = text[:, 0:2]
labels = text[:, 2:]
```

Plot the input data:

```
# Plot input data
plt.figure()
plt.scatter(data[:,0], data[:,1])
plt.xlabel('Dimension 1')
plt.ylabel('Dimension 2')
plt.title('Input data')
```

Extract the minimum and maximum values for each dimension (we don't need to hardcode it like we did in the previous section):

```
# Minimum and maximum values for each dimension
dim1_min, dim1_max = data[:,0].min(), data[:,0].max()
dim2_min, dim2_max = data[:,1].min(), data[:,1].max()
```

Define the number of neurons in the output layer:

```
# Define the number of neurons in the output layer
num_output = labels.shape[1]
```

Define a single layer neural network using the above parameters:

```
# Define a single-layer neural network
dim1 = [dim1_min, dim1_max]
dim2 = [dim2_min, dim2_max]
nn = nl.net.newp([dim1, dim2], num_output)
```

Train the neural network using training data:

```
# Train the neural network
error_progress = nn.train(data, labels, epochs=100, show=20, lr=0.03)
```

Plot the training progress:

```
# Plot the training progress
plt.figure()
plt.plot(error_progress)
plt.xlabel('Number of epochs')
plt.ylabel('Training error')
plt.title('Training error progress')
plt.grid()

plt.show()
```

Define some sample test datapoints and run the network on those points:

```
# Run the classifier on test datapoints
print('\nTest results:')
data_test = [[0.4, 4.3], [4.4, 0.6], [4.7, 8.1]]
for item in data_test:
    print(item, '-->', nn.sim([item])[0])
```

The full code is given in the file `simple_neural_network.py`. If you run the code, you will get two screenshot. The first screenshot represents the input datapoints:

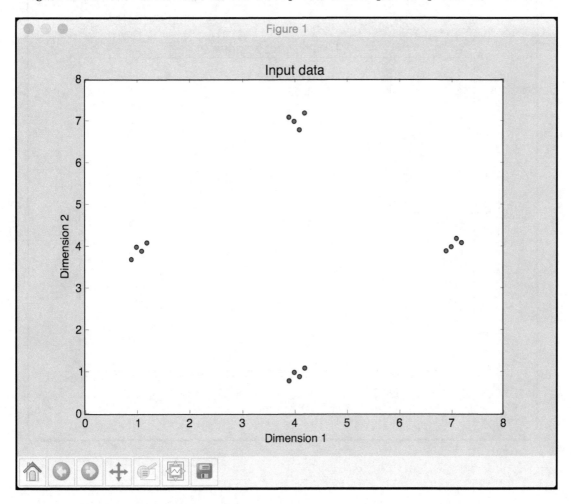

The second screenshot shows the training progress:

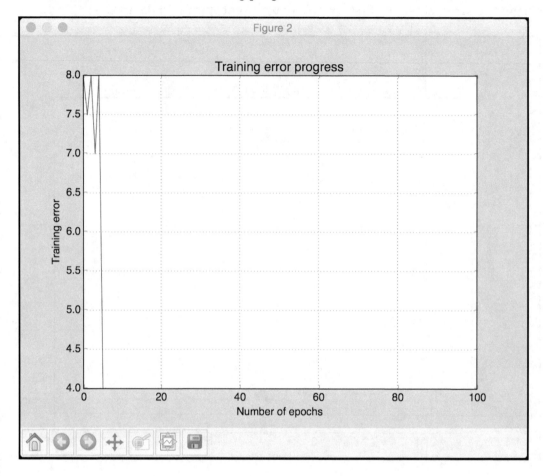

Once you close the graphs, you will see the following printed on your Terminal:

```
Epoch: 20; Error: 4.0;
Epoch: 40; Error: 4.0;
Epoch: 60; Error: 4.0;
Epoch: 80; Error: 4.0;
Epoch: 100; Error: 4.0;
The maximum number of train epochs is reached

Test results:
[0.4, 4.3] --> [ 0.  0.]
[4.4, 0.6] --> [ 1.  0.]
[4.7, 8.1] --> [ 1.  1.]
```

If you locate these test data points on a 2D graph, you can visually verify that the predicted outputs are correct.

Constructing a multilayer neural network

In order to enable higher accuracy, we need to give more freedom to the neural network. This means that a neural network needs more than one layer to extract the underlying patterns in the training data. Let's create a multilayer neural network to achieve that.

Create a new Python file and import the following packages:

```
import numpy as np
import matplotlib.pyplot as plt
import neurolab as nl
```

In the previous two sections, we saw how to use a neural network as a classifier. In this section, we will see how to use a multilayer neural network as a regressor. Generate some sample data points based on the equation $y = 3x^2 + 5$ and then normalize the points:

```
# Generate some training data
min_val = -15
max_val = 15
num_points = 130
x = np.linspace(min_val, max_val, num_points)
y = 3 * np.square(x) + 5
y /= np.linalg.norm(y)
```

Reshape the above variables to create a training dataset:

```
# Create data and labels
data = x.reshape(num_points, 1)
labels = y.reshape(num_points, 1)
```

[373]

Plot the input data:

```
# Plot input data
plt.figure()
plt.scatter(data, labels)
plt.xlabel('Dimension 1')
plt.ylabel('Dimension 2')
plt.title('Input data')
```

Define a multilayer neural network with two hidden layers. You are free to design a neural network any way you want. For this case, let's have 10 neurons in the first layer and 6 neurons in the second layer. Our task is to predict the value, so the output layer will contain a single neuron:

```
# Define a multilayer neural network with 2 hidden layers;
# First hidden layer consists of 10 neurons
# Second hidden layer consists of 6 neurons
# Output layer consists of 1 neuron
nn = nl.net.newff([[min_val, max_val]], [10, 6, 1])
```

Set the training algorithm to gradient descent:

```
# Set the training algorithm to gradient descent
nn.trainf = nl.train.train_gd
```

Train the neural network using the training data that was generated:

```
# Train the neural network
error_progress = nn.train(data, labels, epochs=2000, show=100, goal=0.01)
```

Run the neural network on the training datapoints:

```
# Run the neural network on training datapoints
output = nn.sim(data)
y_pred = output.reshape(num_points)
```

Plot the training progress:

```
# Plot training error
plt.figure()
plt.plot(error_progress)
plt.xlabel('Number of epochs')
plt.ylabel('Error')
plt.title('Training error progress')
```

Plot the predicted output:

```
# Plot the output
x_dense = np.linspace(min_val, max_val, num_points * 2)
y_dense_pred =
nn.sim(x_dense.reshape(x_dense.size,1)).reshape(x_dense.size)

plt.figure()
plt.plot(x_dense, y_dense_pred, '-', x, y, '.', x, y_pred, 'p')
plt.title('Actual vs predicted')

plt.show()
```

The full code is given in the file `multilayer_neural_network.py`. If you run the code, you will get three screenshot. The first screenshot shows the input data:

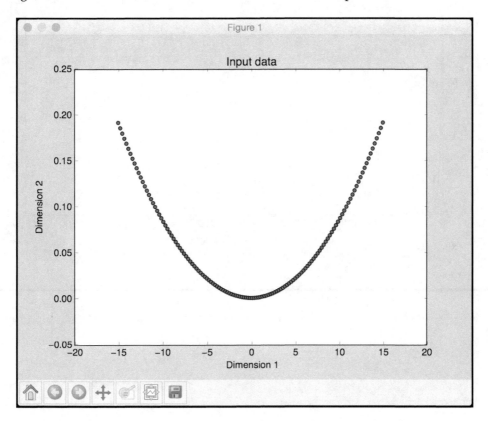

The second screenshot shows the training progress:

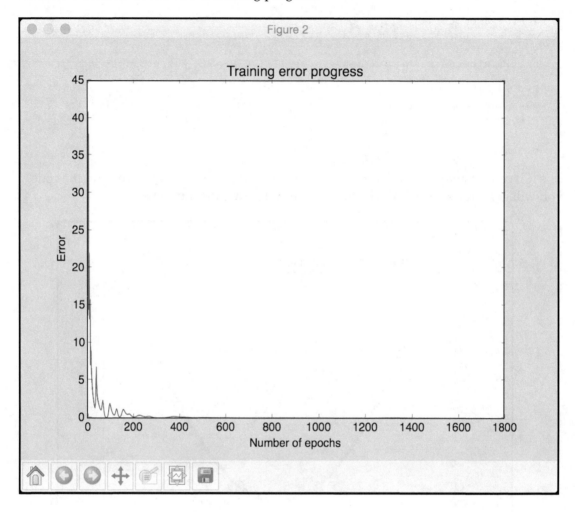

The third screenshot shows the predicted output overlaid on top of input data:

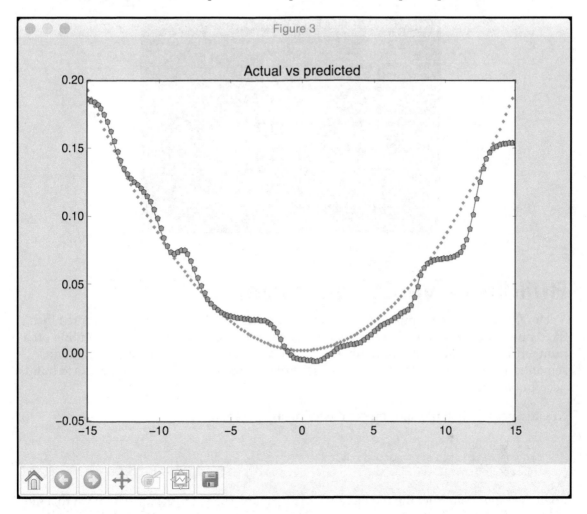

The predicted output seems to follow the general trend. If you continue to train the network and reduce the error, you will see that the predicted output will match the input curve even more accurately.

You will see the following printed on your Terminal:

```
Epoch: 100; Error: 1.9247718251621995;
Epoch: 200; Error: 0.15723294798079526;
Epoch: 300; Error: 0.021680213116912858;
Epoch: 400; Error: 0.1381761995539017;
Epoch: 500; Error: 0.04392553381948737;
Epoch: 600; Error: 0.02975401597014979;
Epoch: 700; Error: 0.014228560930227126;
Epoch: 800; Error: 0.03460207842970052;
Epoch: 900; Error: 0.035934053149433196;
Epoch: 1000; Error: 0.025833284445815966;
Epoch: 1100; Error: 0.013672412879982398;
Epoch: 1200; Error: 0.01776586425692384;
Epoch: 1300; Error: 0.04310242610384976;
Epoch: 1400; Error: 0.03799681296096611;
Epoch: 1500; Error: 0.02467030041520845;
Epoch: 1600; Error: 0.010094873168855236;
Epoch: 1700; Error: 0.01210866043021068;
The goal of learning is reached
```

Building a vector quantizer

Vector Quantization is a quantization technique where the input data is represented by a fixed number of representative points. It is the N-dimensional equivalent of rounding off a number. This technique is commonly used in multiple fields such as image recognition, semantic analysis, and data science. Let's see how to use artificial neural networks to build a vector quantizer.

Create a new Python file and import the following packages:

```
import numpy as np
import matplotlib.pyplot as plt
import neurolab as nl
```

Load the input data from the file `data_vector_quantization.txt`. Each line in this file contains six numbers. The first two numbers form the datapoint and the last four numbers form a one-hot encoded label. There are four classes overall.

```
# Load input data
text = np.loadtxt('data_vector_quantization.txt')
```

Separate the text into data and labels:

```
# Separate it into data and labels
data = text[:, 0:2]
labels = text[:, 2:]
```

Define a neural network with two layers where we have 10 neurons in the input layer and 4 neurons in the output layer:

```
# Define a neural network with 2 layers:
# 10 neurons in input layer and 4 neurons in output layer
num_input_neurons = 10
num_output_neurons = 4
weights = [1/num_output_neurons] * num_output_neurons
nn = nl.net.newlvq(nl.tool.minmax(data), num_input_neurons, weights)
```

Train the neural network using the training data:

```
# Train the neural network
_ = nn.train(data, labels, epochs=500, goal=-1)
```

In order to visualize the output clusters, let's create a grid of points:

```
# Create the input grid
xx, yy = np.meshgrid(np.arange(0, 10, 0.2), np.arange(0, 10, 0.2))
xx.shape = xx.size, 1
yy.shape = yy.size, 1
grid_xy = np.concatenate((xx, yy), axis=1)
```

Evaluate the grid of points using the neural network:

```
# Evaluate the input grid of points
grid_eval = nn.sim(grid_xy)
```

Extract the four classes:

```
# Define the 4 classes
class_1 = data[labels[:,0] == 1]
class_2 = data[labels[:,1] == 1]
class_3 = data[labels[:,2] == 1]
class_4 = data[labels[:,3] == 1]
```

Extract the grids corresponding to those four classes:

```
# Define X-Y grids for all the 4 classes
grid_1 = grid_xy[grid_eval[:,0] == 1]
grid_2 = grid_xy[grid_eval[:,1] == 1]
grid_3 = grid_xy[grid_eval[:,2] == 1]
grid_4 = grid_xy[grid_eval[:,3] == 1]
```

Plot the outputs:

```
# Plot the outputs
plt.plot(class_1[:,0], class_1[:,1], 'ko',
         class_2[:,0], class_2[:,1], 'ko',
```

Artificial Neural Networks

```
        class_3[:,0], class_3[:,1], 'ko',
        class_4[:,0], class_4[:,1], 'ko')
plt.plot(grid_1[:,0], grid_1[:,1], 'm.',
        grid_2[:,0], grid_2[:,1], 'bx',
        grid_3[:,0], grid_3[:,1], 'c^',
        grid_4[:,0], grid_4[:,1], 'y+')
plt.axis([0, 10, 0, 10])
plt.xlabel('Dimension 1')
plt.ylabel('Dimension 2')
plt.title('Vector quantization')

plt.show()
```

The full code is given in the file `vector_quantizer.py`. If you run the code, you will get the following sccreenshot that shows the input data points and the boundaries between clusters:

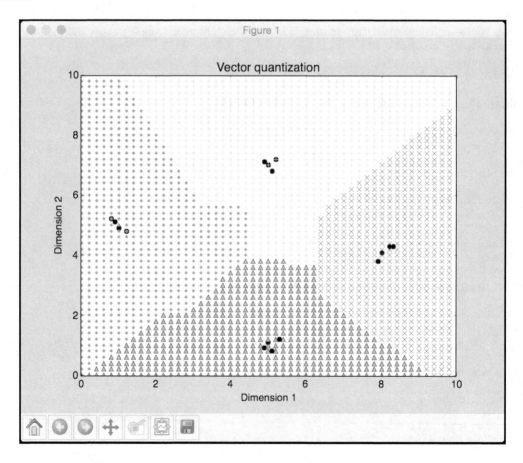

You will see the following printed on your Terminal:

```
Epoch: 100; Error: 0.0;
Epoch: 200; Error: 0.0;
Epoch: 300; Error: 0.0;
Epoch: 400; Error: 0.0;
Epoch: 500; Error: 0.0;
The maximum number of train epochs is reached
```

Analyzing sequential data using recurrent neural networks

We have been dealing with static data so far. Artificial neural networks are good at building models for sequential data too. In particular, recurrent neural networks are great at modeling sequential data. Perhaps time-series data is the most commonly occurring form of sequential data in our world. You can learn more about recurrent neural networks at http://www.wildml.com/2015/09/recurrent-neural-networks-tutorial-part-1-introduction-to-rnns. When we are working with time-series data, we cannot just use generic learning models. We need to characterize the temporal dependencies in our data so that we can build a robust model. Let's see how to build it.

Create a new python file and import the following packages:

```
import numpy as np
import matplotlib.pyplot as plt
import neurolab as nl
```

Define a function to generate the waveforms. Start by defining four sine waves:

```
def get_data(num_points):
    # Create sine waveforms
    wave_1 = 0.5 * np.sin(np.arange(0, num_points))
    wave_2 = 3.6 * np.sin(np.arange(0, num_points))
    wave_3 = 1.1 * np.sin(np.arange(0, num_points))
    wave_4 = 4.7 * np.sin(np.arange(0, num_points))
```

Create varying amplitudes for the overall waveform:

```
    # Create varying amplitudes
    amp_1 = np.ones(num_points)
    amp_2 = 2.1 + np.zeros(num_points)
    amp_3 = 3.2 * np.ones(num_points)
    amp_4 = 0.8 + np.zeros(num_points)
```

Create the overall waveform:

```
    wave = np.array([wave_1, wave_2, wave_3, wave_4]).reshape(num_points *
4, 1)
    amp = np.array([[amp_1, amp_2, amp_3, amp_4]]).reshape(num_points * 4,
1)

    return wave, amp
```

Define a function to visualize the output of the neural network:

```
# Visualize the output
def visualize_output(nn, num_points_test):
    wave, amp = get_data(num_points_test)
    output = nn.sim(wave)
    plt.plot(amp.reshape(num_points_test * 4))
    plt.plot(output.reshape(num_points_test * 4))
```

Define the main function and create a waveform:

```
if __name__=='__main__':
    # Create some sample data
    num_points = 40
    wave, amp = get_data(num_points)
```

Create a recurrent neural network with two layers:

```
    # Create a recurrent neural network with 2 layers
    nn = nl.net.newelm([[-2, 2]], [10, 1], [nl.trans.TanSig(),
nl.trans.PureLin()])
```

Set the initializer functions for each layer:

```
    # Set the init functions for each layer
    nn.layers[0].initf = nl.init.InitRand([-0.1, 0.1], 'wb')
    nn.layers[1].initf = nl.init.InitRand([-0.1, 0.1], 'wb')
    nn.init()
```

Train the neural network:

```
# Train the recurrent neural network
error_progress = nn.train(wave, amp, epochs=1200, show=100, goal=0.01)
```

Run the data through the network:

```
# Run the training data through the network
output = nn.sim(wave)
```

Plot the output:

```
# Plot the results
plt.subplot(211)
plt.plot(error_progress)
plt.xlabel('Number of epochs')
plt.ylabel('Error (MSE)')

plt.subplot(212)
plt.plot(amp.reshape(num_points * 4))
plt.plot(output.reshape(num_points * 4))
plt.legend(['Original', 'Predicted'])
```

Test the performance of the neural network on unknown test data:

```
# Testing the network performance on unknown data
plt.figure()

plt.subplot(211)
visualize_output(nn, 82)
plt.xlim([0, 300])

plt.subplot(212)
visualize_output(nn, 49)
plt.xlim([0, 300])

plt.show()
```

Artificial Neural Networks

The full code is given in the file `recurrent_neural_network.py`. If you run the code, you will see two output figures. The upper half of the first screenshot shows the training progress and the lower half shows the predicted output overlaid on top of the input waveform:

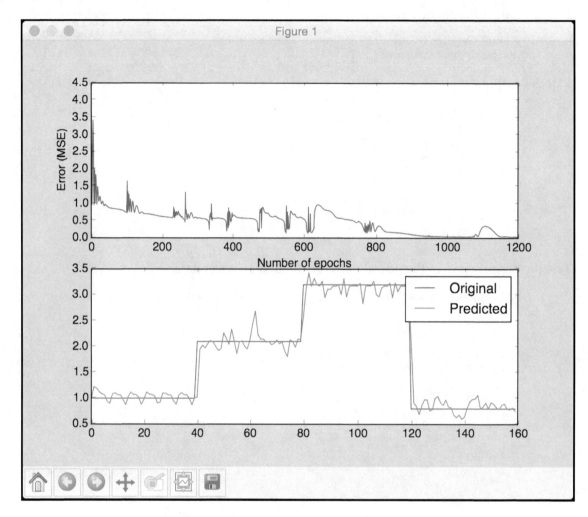

The upper half of the second screenshot shows how the neural network simulates the waveform even though we increase the length of the waveform. The lower half of the screenshot shows the same for decreased length.

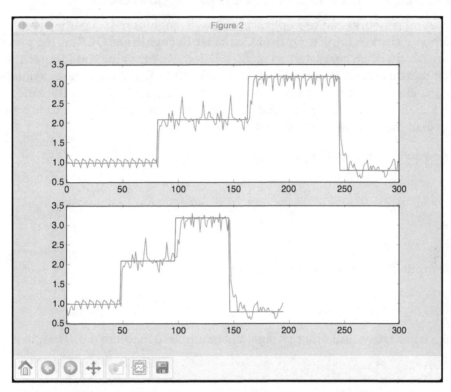

You will see the following printed on your Terminal:

```
Epoch: 100; Error: 0.7378753203612153;
Epoch: 200; Error: 0.6276459886666788;
Epoch: 300; Error: 0.586316536629095;
Epoch: 400; Error: 0.7246461052491963;
Epoch: 500; Error: 0.7244266943409208;
Epoch: 600; Error: 0.5650581389122635;
Epoch: 700; Error: 0.5798180931911314;
Epoch: 800; Error: 0.19557566610789826;
Epoch: 900; Error: 0.10837074465396046;
Epoch: 1000; Error: 0.04330852391940663;
Epoch: 1100; Error: 0.3073835343028226;
Epoch: 1200; Error: 0.034685278416163604;
The maximum number of train epochs is reached
```

Visualizing characters in an Optical Character Recognition database

Artificial neural networks can use optical character recognition. It is perhaps one of the most commonly sited examples. **Optical Character Recognition (OCR)** is the process of recognizing handwritten characters in images. Before we jump into building that model, we need to familiarize ourselves with the dataset. We will be using the dataset available at http://ai.stanford.edu/~btaskar/ocr. You will be downloading a file called letter.data. For convenience, this file has been provided to you in the code bundle. Let's see how to load that data and visualize the characters.

Create a new python file and import the following packages:

```
import os
import sys

import cv2
import numpy as np
```

Define the input file containing the OCR data:

```
# Define the input file
input_file = 'letter.data'
```

Define the visualization and other parameters required to load the data from that file:

```
# Define the visualization parameters
img_resize_factor = 12
start = 6
end = -1
height, width = 16, 8
```

Iterate through the lines of that file until the user presses the Esc key. Each line in that file is tab separated. Read each line and scale it up to 255:

```
# Iterate until the user presses the Esc key
with open(input_file, 'r') as f:
    for line in f.readlines():
        # Read the data
        data = np.array([255 * float(x) for x in line.split('\t')[start:end]])
```

Reshape the 1D array into a 2D image:

```
        # Reshape the data into a 2D image
        img = np.reshape(data, (height, width))
```

Scale the image for visualization:

```
# Scale the image
img_scaled = cv2.resize(img, None, fx=img_resize_factor, fy=img_resize_factor)
```

Display the image:

```
# Display the image
cv2.imshow('Image', img_scaled)
```

Check if the user has pressed the Esc key. If so, exit the loop:

```
# Check if the user pressed the Esc key
c = cv2.waitKey()
if c == 27:
    break
```

The full code is given in the file `character_visualizer.py`. If you run the code, you will get an output screenshot displaying a character. You can keep pressing the space bar to see more characters. An o looks like this:

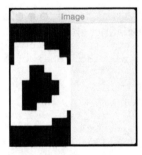

An i looks like this:

Building an Optical Character Recognition engine

Now that we have learned how to work with this data, let's build an optical character recognition system using artificial neural networks.

Create a new python file and import the following packages:

```
import numpy as np
import neurolab as nl
```

Define the input file:

```
# Define the input file
input_file = 'letter.data'
```

Define the number of datapoints that will be loaded:

```
# Define the number of datapoints to
# be loaded from the input file
num_datapoints = 50
```

Define the string containing all the distinct characters:

```
# String containing all the distinct characters
orig_labels = 'omandig'
```

Extract the number of distinct classes:

```
# Compute the number of distinct characters
num_orig_labels = len(orig_labels)
```

Define the train and test split. We will use 90% for training and 10% for testing:

```
# Define the training and testing parameters
num_train = int(0.9 * num_datapoints)
num_test = num_datapoints - num_train
```

Define the dataset extraction parameters:

```
# Define the dataset extraction parameters
start = 6
end = -1
```

Create the dataset:

```
# Creating the dataset
data = []
labels = []
with open(input_file, 'r') as f:
    for line in f.readlines():
        # Split the current line tabwise
        list_vals = line.split('\t')
```

If the label is not in our list of labels, we should skip it:

```
        # Check if the label is in our ground truth
        # labels. If not, we should skip it.
        if list_vals[1] not in orig_labels:
            continue
```

Extract the current label and append it to the main list:

```
        # Extract the current label and append it
        # to the main list
        label = np.zeros((num_orig_labels, 1))
        label[orig_labels.index(list_vals[1])] = 1
        labels.append(label)
```

Extract the character vector and append it to the main list:

```
        # Extract the character vector and append it to the main list
        cur_char = np.array([float(x) for x in list_vals[start:end]])
        data.append(cur_char)
```

Exit the loop once we have created the dataset:

```
        # Exit the loop once the required dataset has been created
        if len(data) >= num_datapoints:
            break
```

Convert the lists into numpy arrays:

```
# Convert the data and labels to numpy arrays
data = np.asfarray(data)
labels = np.array(labels).reshape(num_datapoints, num_orig_labels)
```

Extract the number of dimensions:

```
# Extract the number of dimensions
num_dims = len(data[0])
```

Create a feedforward neural network and set the training algorithm to gradient descent:

```
# Create a feedforward neural network
nn = nl.net.newff([[0, 1] for _ in range(len(data[0]))],
        [128, 16, num_orig_labels])

# Set the training algorithm to gradient descent
nn.trainf = nl.train.train_gd
```

Train the neural network:

```
# Train the network
error_progress = nn.train(data[:num_train,:], labels[:num_train,:],
        epochs=10000, show=100, goal=0.01)
```

Predict the output for test data:

```
# Predict the output for test inputs
print('\nTesting on unknown data:')
predicted_test = nn.sim(data[num_train:, :])
for i in range(num_test):
    print('\nOriginal:', orig_labels[np.argmax(labels[i])])
    print('Predicted:', orig_labels[np.argmax(predicted_test[i])])
```

The full code is given in the file `ocr.py`. If you run the code, you will see the following on your Terminal:

```
Epoch: 100; Error: 80.75182001223291;
Epoch: 200; Error: 49.823887961230206;
Epoch: 300; Error: 26.624261963923217;
Epoch: 400; Error: 31.131906412329677;
Epoch: 500; Error: 30.589610928772494;
Epoch: 600; Error: 23.129959531324324;
Epoch: 700; Error: 15.561849160600984;
Epoch: 800; Error: 9.52433563455828;
Epoch: 900; Error: 1.4032941634688987;
Epoch: 1000; Error: 1.1584148924740179;
Epoch: 1100; Error: 0.844934060039839;
Epoch: 1200; Error: 0.646187646028962;
Epoch: 1300; Error: 0.48881681329304894;
Epoch: 1400; Error: 0.4005475591737743;
Epoch: 1500; Error: 0.34145887283532067;
Epoch: 1600; Error: 0.29871068426249625;
Epoch: 1700; Error: 0.2657577763744411;
Epoch: 1800; Error: 0.23921810237252988;
Epoch: 1900; Error: 0.2172060084455509;
Epoch: 2000; Error: 0.19856823374761018;
Epoch: 2100; Error: 0.18253521958793384;
Epoch: 2200; Error: 0.16855895648078095;
```

It will keep going until 10,000 epochs. Once it's done, you will see the following on your Terminal:

```
Epoch: 9500; Error: 0.032460181065798295;
Epoch: 9600; Error: 0.027044816600106478;
Epoch: 9700; Error: 0.022026328910164213;
Epoch: 9800; Error: 0.018353324233938713;
Epoch: 9900; Error: 0.015789692591136868;
Epoch: 10000; Error: 0.014064205770213847;
The maximum number of train epochs is reached

Testing on unknown data:

Original: o
Predicted: o

Original: m
Predicted: n

Original: m
Predicted: m

Original: a
Predicted: d

Original: n
Predicted: n
```

As we can see in the preceding screenshot, it gets three of them right. If you use a bigger dataset and train longer, then you will get higher accuracy.

Summary

In this chapter, we learned about artificial neural networks. We discussed how to build and train neural networks. We talked about perceptrons and built a classifier based on that. We learned about single layer neural networks as well as multilayer neural networks. We discussed how neural networks could be used to build a vector quantizer. We analyzed sequential data using recurrent neural networks. We then built an optical character recognition engine using artificial neural networks. In the next chapter, we will learn about reinforcement learning and see how to build smart learning agents.

15
Reinforcement Learning

In this chapter, we are going to learn about reinforcement learning. We will discuss the premise of reinforcement learning. We will talk about the differences between reinforcement learning and supervised learning. We will go through some real world examples of reinforcement learning and see how it manifests itself in various forms. We will learn about the building blocks of reinforcement learning and the various concepts involved. We will then create an environment in python to see how it works in practice. We will then use these concepts to build a learning agent.

By the end of this chapter, you will know:

- Understanding the premise
- Reinforcement learning vs. supervised learning
- Real world examples of reinforcement learning
- Building blocks of reinforcement learning
- Creating an environment
- Building a learning agent

Understanding the premise

The concept of learning is fundamental to Artificial Intelligence. We want the machines to understand the process of learning so that they can do it on their own. Humans learn by observing and interacting with their surroundings. When you go to a new place, you quickly scan and see what's happening around you. Nobody is teaching you what to do here. You are observing and interacting with the environment around you. By building this connection with the environment, we tend to gather a lot of information about what's causing different things. We learn about cause and effect, what actions lead to what results, and what we need to do in order to achieve something.

We use this premise everywhere in our lives. We gather all this knowledge about our surroundings and, in turn, learn how we respond to that. Let's consider another example of an orator. Whenever good orators are giving speeches in public, they are aware of how the crowd is reacting to what they are saying. If the crowd is not responding to it, then the orator changes the speech in real time to ensure that the crowd is engaged. As we can see, the orator is trying to influence the environment through his/her behavior. We can say that the orator *learned* from interaction with the crowd in order to take action to achieve a certain *goal*. This is one of the most fundamental ideas in Artificial Intelligence on which many topics are based. Let's talk about reinforcement learning by keeping this in mind.

Reinforcement learning refers to the process of learning what to do and mapping situations to certain actions in order to maximize the reward. In most paradigms of machine learning, a learning agent is told what actions to take in order to achieve certain results. In the case of reinforcement leaning, the learning agent is not told what actions to take. Instead, it must discover what actions yield the highest reward by trying them out. These actions tend to affect the immediate reward as well as the next situation. This means that all the subsequent rewards will be affected too.

A good way to think about reinforcement learning is by understanding that we are defining a learning problem and not a learning method. So we can say that any method that can solve our problem can be considered as a reinforcement learning method. Reinforcement learning is characterized by two distinguishing features — trial and error learning, and delayed reward. A reinforcement learning agent uses these two features to learn from the consequences of its actions.

Reinforcement learning versus supervised learning

A lot of current research is focused on supervised learning. Reinforcement learning might seem a bit similar to supervised learning, but it is not. The process of supervised learning refers to learning from labeled samples provided by us. While this is a very useful technique, it is not sufficient to start learning from interactions. When we want to design a machine to navigate unknown terrains, this kind of learning is not going to help us. We don't have training samples available beforehand. We need an agent that can learn from its own experience by interacting with the unknown terrain. This is where reinforcement learning really shines.

Let's consider the exploration part where the agent has to interact with the new environment in order to learn. How much can it possibly explore? We do not even know how big the environment is, and in most cases, it is not possible to explore all the possibilities. So what should the agent do? Should it learn from its limited experience or wait until it explores further before taking action? This is one of the main challenges of reinforcement learning. In order to get a higher reward, an agent must favor the actions that have been tried and tested. But in order to discover such actions, it has to keep trying newer actions that have not been selected before. Researchers have studied this trade off between exploration and exploitation extensively over the years and it's still an active topic.

Real world examples of reinforcement learning

Let's see where reinforcement learning occurs in the real world. This will help us understand how it works and what possible applications can be built using this concept:

- **Game playing**: Let's consider a board game like **Go** or **Chess.** In order to determine the best move, the players need to think about various factors. The number of possibilities is so large that it is not possible to perform a brute-force search. If we were to build a machine to play such a game using traditional techniques, we need to specify a large number of rules to cover all these possibilities. Reinforcement learning completely bypasses this problem. We do not need to manually specify any rules. The learning agent simply learns by actually playing the game.
- **Robotics**: Let's consider a robot whose job is to explore a new building. It has to make sure it has enough power left to come back to the base station. This robot has to decide if it should make decisions by considering the trade off between the amount of information collected and the ability to reach back to base station safely.
- **Industrial controllers:** Consider the case of scheduling elevators. A good scheduler will spend the least amount of power and service the highest number of people. For problems like these, reinforcement learning agents can learn how to do this in a simulated environment. They can then take that knowledge to come up with optimal scheduling.
- **Babies:** Newborns struggle to walk in the first few months. They learn by trying it over and over again until they learn how to balance.

If you observe these examples closely, you will see there are some common traits. All of them involve interacting with the environment. The learning agent aims to achieve a certain goal even though there's uncertainty about the environment. The actions of an agent will change the future state of that environment. This impacts the opportunities available at later times as the agent continues to interact with the environment.

Building blocks of reinforcement learning

Now that we have seen a few examples, let's dig into the building blocks of a reinforcement learning system. Apart from the interaction between the agent and the environment, there are other factors at play here:

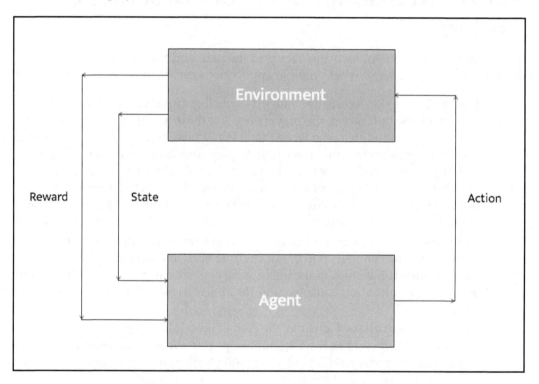

A typical reinforcement learning agent goes through the following steps:

- There is a set of states related to the agent and the environment. At a given point of time, the agent observes an input state to sense the environment.
- There are policies that govern what action needs to be taken. These policies act as decision making functions. The action is determined based on the input state using these policies.
- The agent takes the action based on the previous step.
- The environment reacts in a particular way in response to that action. The agent receives reinforcement, also known as reward, from the environment.
- The agent records the information about this reward. It's important to note that this reward is for this particular pair of state and action.

Reinforcement learning systems can do multiple things simultaneously — learn by performing a trial and error search, learn the model of the environment it is in, and then use that model to plan the next steps.

Creating an environment

We will be using a package called `OpenAI Gym` to build reinforcement learning agents. You can learn more about it here: `https://gym.openai.com`. We can install it using pip by running the following command on the Terminal:

```
$ pip3 install gym
```

You can find various tips and tricks related to its installation here: `https://github.com/openai/gym#installation`. Now that you have installed it, let's go ahead and write some code.

Create a new python file and import the following package:

```
import argparse

import gym
```

Define a function to parse the input arguments. We will be able to specify the type of environment we want to run:

```
def build_arg_parser():
    parser = argparse.ArgumentParser(description='Run an environment')
    parser.add_argument('--input-env', dest='input_env', required=True,
            choices=['cartpole', 'mountaincar', 'pendulum', 'taxi', 'lake'],
            help='Specify the name of the environment')
    return parser
```

Define the main function and parse the input arguments:

```
if __name__=='__main__':
    args = build_arg_parser().parse_args()
    input_env = args.input_env
```

Create a mapping from input argument string to the names of the environments as specified in the `OpenAI Gym` package:

```
    name_map = {'cartpole': 'CartPole-v0',
                'mountaincar': 'MountainCar-v0',
                'pendulum': 'Pendulum-v0',
                'taxi': 'Taxi-v1',
                'lake': 'FrozenLake-v0'}
```

Create the environment based on the input argument and reset it:

```
    # Create the environment and reset it
    env = gym.make(name_map[input_env])
    env.reset()
```

Iterate 1000 times and take action during each step:

```
    # Iterate 1000 times
    for _ in range(1000):
        # Render the environment
        env.render()

        # take a random action
        env.step(env.action_space.sample())
```

The full code is given in the file `run_environment.py`. If you want to know how to run the code, run it with the help argument as shown in the following figure:

```
$ python3 run_environment.py --help
usage: run_environment.py [-h] --input-env
                          {cartpole,mountaincar,pendulum,taxi,lake}

Run an environment

optional arguments:
  -h, --help            show this help message and exit
  --input-env {cartpole,mountaincar,pendulum,taxi,lake}
                        Specify the name of the environment
```

Let's run it with the cartpole environment. Run the following command on your Terminal:

```
$ python3 run_environment.py --input-env cartpole
```

If you run it, you will see a window showing a **cartpole** moving to your right. The following screenshot shows the initial position:

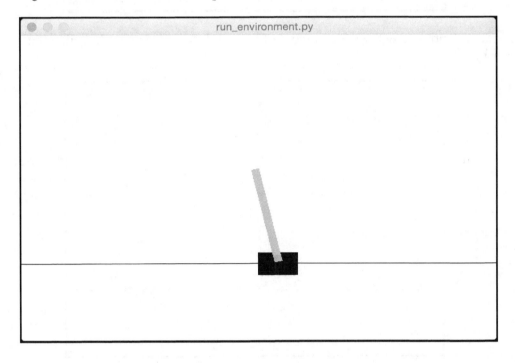

In the next second or so, you will see it moving as shown in the following screenshot:

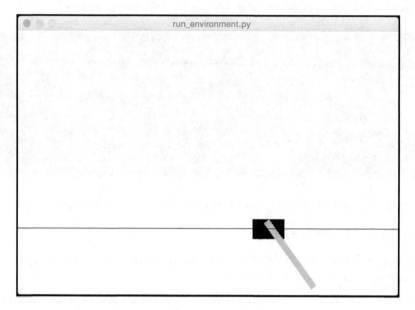

Towards the end, you will see it going out of the window as shown in the following screenshot:

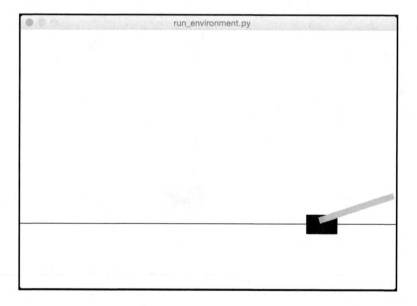

Let's run it with the mountain car argument. Run the following command on your Terminal:

```
$ python3 run_environment.py --input-env mountaincar
```

If you run the code, you will see the following figure initially:

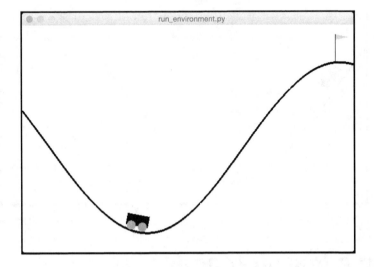

If you let it run for a few seconds, you will see that the car oscillates more in order to reach the flag:

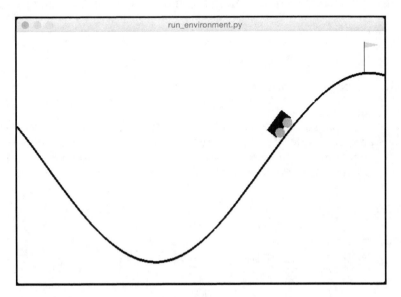

It will keep taking longer strides as shown in the following figure:

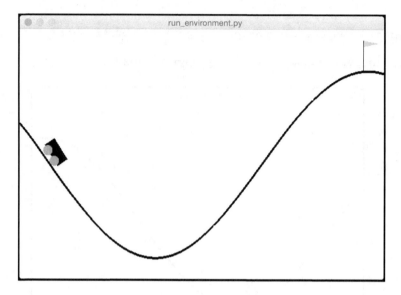

Building a learning agent

Let's see how to build a learning agent that can achieve a goal. The learning agent will learn how to achieve a goal. Create a new python file and import the following package:

```
import argparse

import gym
```

Define a function to parse the input arguments:

```
def build_arg_parser():
    parser = argparse.ArgumentParser(description='Run an environment')
    parser.add_argument('--input-env', dest='input_env', required=True,
            choices=['cartpole', 'mountaincar', 'pendulum'],
            help='Specify the name of the environment')
    return parser
```

Parse the input arguments:

```
if __name__=='__main__':
    args = build_arg_parser().parse_args()
    input_env = args.input_env
```

Build a mapping from the input arguments to the names of the environments in the OpenAI Gym package:

```
name_map = {'cartpole': 'CartPole-v0',
            'mountaincar': 'MountainCar-v0',
            'pendulum': 'Pendulum-v0'}
```

Create the environment based on the input argument:

```
# Create the environment
env = gym.make(name_map[input_env])
```

Start iterating by resetting the environment:

```
# Start iterating
for _ in range(20):
    # Reset the environment
    observation = env.reset()
```

For each reset, iterate 100 times. Start by rendering the environment:

```
    # Iterate 100 times
    for i in range(100):
        # Render the environment
        env.render()
```

Print the current observation and take action based on the available action space:

```
        # Print the current observation
        print(observation)

        # Take action
        action = env.action_space.sample()
```

Extract the consequences of taking the current action:

```
        # Extract the observation, reward, status and
        # other info based on the action taken
        observation, reward, done, info = env.step(action)
```

Check if we have achieved our goal:

```
        # Check if it's done
        if done:
            print('Episode finished after {} timesteps'.format(i+1))
            break
```

Reinforcement Learning

The full code is given in the file `balancer.py`. If you want to know how to run the code, run it with the help argument as shown in the following screenshot:

```
$ python3 balancer.py --help
usage: balancer.py [-h] --input-env {cartpole,mountaincar,pendulum}

Run an environment

optional arguments:
  -h, --help            show this help message and exit
  --input-env {cartpole,mountaincar,pendulum}
                        Specify the name of the environment
```

Let's run the code with the `cartpole` environment. Run the following command on your Terminal:

```
$ python3 balancer.py --input-env cartpole
```

If you run the code, you will see that the cartpole balances itself:

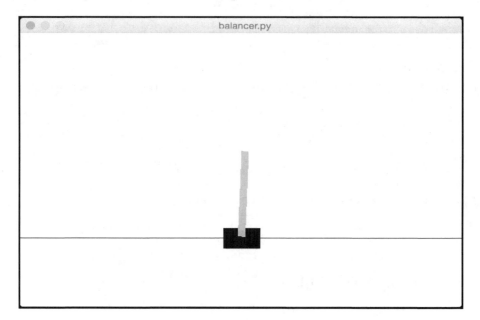

If you let it run for a few seconds, you will see that the cartpole is still standing as shown in the following screenshot:

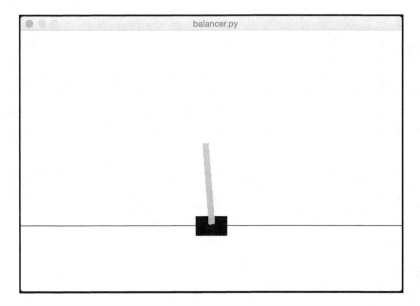

You will see a lot of information printed on your Terminal. If you look at one of the episodes, it will look something like this:

```
[ 0.01704777  0.03379922 -0.01628054  0.02868271]
[ 0.01772375 -0.16108552 -0.01570689  0.31618481]
[ 0.01450204  0.03425659 -0.00938319  0.01859014]
[ 0.01518717 -0.16072954 -0.00901139  0.30829785]
[ 0.01197258 -0.35572194 -0.00284543  0.59812526]
[ 0.00485814 -0.16056029  0.00911707  0.30454742]
[ 0.00164694 -0.35581098  0.01520802  0.60009165]
[-0.00546928 -0.16090505  0.02720986  0.31223756]
[-0.00868738 -0.35640386  0.03345461  0.61337594]
[-0.01581546 -0.55197696  0.04572213  0.91640525]
[-0.026855   -0.3575021   0.06405023  0.63843544]
[-0.03400504 -0.16332896  0.07681894  0.36659087]
[-0.03727162 -0.3594537   0.08415076  0.68247294]
[-0.04446069 -0.5556372   0.09780022  1.00041801]
[-0.05557344 -0.75192055  0.11780858  1.32214352]
[-0.07061185 -0.55846765  0.14425145  1.06853119]
[-0.0817812  -0.36551752  0.16562207  0.82437502]
[-0.08909155 -0.56247052  0.18210957  1.16423244]
[-0.10034096 -0.75943464  0.20539422  1.50803784]
Episode finished after 19 timesteps
```

Different episodes take a different number of steps to finish. If you scroll through the information printed on your Terminal, you will be able to see that.

Summary

In this chapter, we learnt about reinforcement learning systems. We discussed the premise of reinforcement learning and how we can set it up. We talked about the differences between reinforcement learning and supervised learning. We went through some real world examples of reinforcement learning and saw how various systems use it in different forms.

We discussed the building blocks of reinforcement learning and concepts such as agent, environment, policy, reward, and so on. We then created an environment in python to see it in action. We used these concepts to build a reinforcement learning agent.

16
Deep Learning with Convolutional Neural Networks

In this chapter, we are going to learn about Deep Learning and **Convolutional Neural Networks** (**CNNs**). CNNs have gained a lot of momentum over the last few years, especially in the field of image recognition. We will talk about the architecture of CNNs and the type of layers used inside. We are going to see how to use a package called `TensorFlow`. We will build a perceptron based linear regressor. We are going to learn how to build an image classifier using a single layer neural network. We will then build an image classifier using a CNN.

By the end of this chapter, you will know:

- What are Convolutional Neural Networks (CNNs)?
- The architecture of CNNs
- The types of layers in a CNN
- Building a perceptron based linear regressor
- Building an image classifier using a single layer neural network
- Building an image classifier using a Convolutional Neural Network

What are Convolutional Neural Networks?

We saw how neural networks work in the last two chapters. Neural networks consist of neurons that have weights and biases. These weights and biases are tuned during the training process to come up with a good learning model. Each neuron receives a set of inputs, processes it in some way, and then outputs a value.

If we build a neural network with many layers, it's called a deep neural network. The branch of Artificial Intelligence dealing with these deep neural networks is referred to as deep learning.

One of the main disadvantages of ordinary neural networks is that they ignore the structure of input data. All data is converted to a single dimensional array before feeding it into the network. This works well for regular data, but things get difficult when we deal with images.

Let's consider grayscale images. These images are 2D structures and we know that the spatial arrangement of pixels has a lot of hidden information. If we ignore this information, we will be losing a lot of underlying patterns. This is where **Convolutional Neural Networks (CNNs)** come into the picture. CNNs take the 2D structure of the images into account when they process them.

CNNs are also made up of neurons consisting of weights and biases. These neurons accept input data, process it, and then output something. The goal of the network is to go from the raw image data in the input layer to the correct class in the output layer. The difference between ordinary neural networks and CNNs is in the type of layers we use and how we treat the input data. CNNs assume that the inputs are images, which allows them to extract properties specific to images. This makes CNNs way more efficient in dealing with images. Let's see how CNNs are built.

Architecture of CNNs

When we are working with ordinary neural networks, we need to convert the input data into a single vector. This vector acts as the input to the neural network, which then passes through the layers of the neural network. In these layers, each neuron is connected to all the neurons in the previous layer. It is also worth noting that the neurons within each layer are not connected to each other. They are only connected to the neurons in the adjacent layers. The last layer in the network is the output layer and it represents the final output.

If we use this structure for images, it will quickly become unmanageable. For example, let's consider an image dataset consisting of *256×256* RGB images. Since these are 3 channel images, there would be *256 * 256 * 3 = 196,608* weights. Note that this is just for a single neuron! Each layer will have multiple neurons, so the number of weights tends to increase rapidly. This means that the model will now have an enormous number of parameters to tune during the training process. This is why it becomes very complex and time-consuming. Connecting each neuron to every neuron in the previous layer, called full connectivity, is clearly not going to work for us.

CNNs explicitly consider the structure of images when processing the data. The neurons in CNNs are arranged in 3 dimensions — width, height, and depth. Each neuron in the current layer is connected to a small patch of the output from the previous layer. It's like overlaying an *NxN* filter on the input image. This is in contrast to a fully connected layer where each neuron is connected to all the neurons of the previous layer.

Since a single filter cannot capture all the nuances of the image, we do this *M* number of times to make sure we capture all the details. These *M* filters act as feature extractors. If you look at the outputs of these filters, we can see that they extract features like edges, corners, and so on. This is true for the initial layers in the CNN. As we progress through layers of the network, we will see that the later layers extract higher level features.

Types of layers in a CNN

Now that we know about the architecture of a CNN, let's see what type of layers are used to construct it. CNNs typically use the following types of layers:

- **Input layer:** This layer takes the raw image data as it is.
- **Convolutional layer:** This layer computes the convolutions between the neurons and the various patches in the input. If you need a quick refresher on image convolutions, you can check out this link: http://web.pdx.edu/~jduh/courses/Archive/geog481w07/Students/Ludwig_ImageConvolution.pdf. The convolutional layer basically computes the dot product between the weights and a small patch in the output of the previous layer.
- **Rectified Linear Unit layer:** This layer applies an activation function to the output of the previous layer. This function is usually something like *max(0, x)*. This layer is needed to add non-linearity to the network so that it can generalize well to any type of function.
- **Pooling layer:** This layer samples the output of the previous layer resulting in a structure with smaller dimensions. Pooling helps us to keep only the prominent parts as we progress in the network. Max pooling is frequently used in the pooling layer where we pick the maximum value in a given *KxK* window.
- **Fully Connected layer:** This layer computes the output scores in the last layer. The resulting output is of the size *1x1xL*, where *L* is the number of classes in the training dataset.

As we go from the input layer to the output layer in the network, the input image gets transformed from pixel values to the final class scores. Many different architectures for CNNs have been proposed and it's an active area of research. The accuracy and robustness of a model depends on many factors — the type of layers, depth of the network, the arrangement of various types of layers within the network, the functions chosen for each layer, training data, and so on.

Building a perceptron-based linear regressor

We will see how to build a linear regression model using perceptrons. We have already seen linear regression in previous chapters, but this section is about building a linear regression model using a neural network approach.

We will be using `TensorFlow` in this chapter. It is a popular deep learning package that's widely used to build various real world systems. In this section, we will get familiar with how it works. Make sure to install it before you proceed. The installation instructions are given here: https://www.tensorflow.org/get_started/os_setup. Once you verify that it's installed, create a new python and import the following packages:

```
import numpy as np
import matplotlib.pyplot as plt
import tensorflow as tf
```

We will be generating some datapoints and see how we can fit a model to it. Define the number of datapoints to be generated:

```
# Define the number of points to generate
num_points = 1200
```

Define the parameters that will be used to generate the data. We will be using the model of a line: $y = mx + c$:

```
# Generate the data based on equation y = mx + c
data = []
m = 0.2
c = 0.5
for i in range(num_points):
    # Generate 'x'
    x = np.random.normal(0.0, 0.8)
```

Generate some noise to add some variation in the data:

```
# Generate some noise
noise = np.random.normal(0.0, 0.04)
```

Compute the value of y using the equation:

```
# Compute 'y'
y = m*x + c + noise

data.append([x, y])
```

Once you finish iterating, separate the data into input and output variables:

```
# Separate x and y
x_data = [d[0] for d in data]
y_data = [d[1] for d in data]
```

Plot the data:

```
# Plot the generated data
plt.plot(x_data, y_data, 'ro')
plt.title('Input data')
plt.show()
```

Generate weights and biases for the perceptron. For weights, we will use a uniform random number generator and set the biases to zero:

```
# Generate weights and biases
W = tf.Variable(tf.random_uniform([1], -1.0, 1.0))
b = tf.Variable(tf.zeros([1]))
```

Define the equation using `TensorFlow` variables:

```
# Define equation for 'y'
y = W * x_data + b
```

Define the loss function that can be used during the training process. The optimizer will try to minimize this value as much as possible.

```
# Define how to compute the loss
loss = tf.reduce_mean(tf.square(y - y_data))
```

Define the gradient descent optimizer and specify the loss function:

```
# Define the gradient descent optimizer
optimizer = tf.train.GradientDescentOptimizer(0.5)
train = optimizer.minimize(loss)
```

All the variables are in place, but they haven't been initialized yet. Let's do that:

```
# Initialize all the variables
init = tf.initialize_all_variables()
```

Start the TensorFlow session and run it using the initializer:

```
# Start the tensorflow session and run it
sess = tf.Session()
sess.run(init)
```

Start the training process:

```
# Start iterating
num_iterations = 10
for step in range(num_iterations):
    # Run the session
    sess.run(train)
```

Print the progress of the training process. The `loss` parameter will continue to decrease as we go through iterations:

```
    # Print the progress
    print('\nITERATION', step+1)
    print('W =', sess.run(W)[0])
    print('b =', sess.run(b)[0])
    print('loss =', sess.run(loss))
```

Plot the generated data and overlay the predicted model on top. In this case, the model is a line:

```
    # Plot the input data
    plt.plot(x_data, y_data, 'ro')

    # Plot the predicted output line
    plt.plot(x_data, sess.run(W) * x_data + sess.run(b))
```

Set the parameters for the plot:

```
    # Set plotting parameters
    plt.xlabel('Dimension 0')
    plt.ylabel('Dimension 1')
    plt.title('Iteration ' + str(step+1) + ' of ' + str(num_iterations))
    plt.show()
```

The full code is given in the file `linear_regression.py`. If you run the code, you will see following screenshot showing input data:

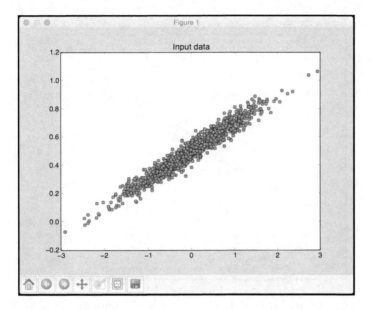

If close you this window, you will see the training process. The first iteration looks like this:

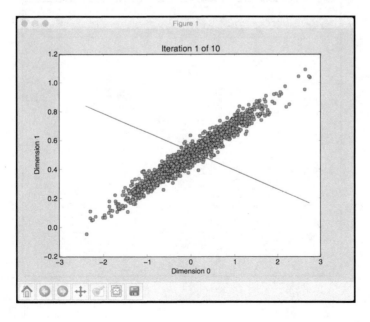

As we can see, the line is completely off. Close this window to go to the next iteration:

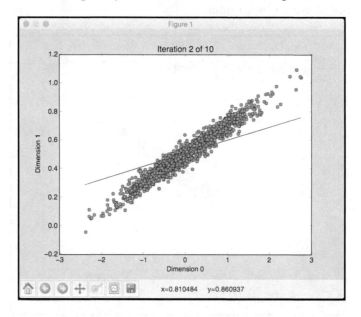

The line seems better, but it's still off. Let's close this window and continue iterating:

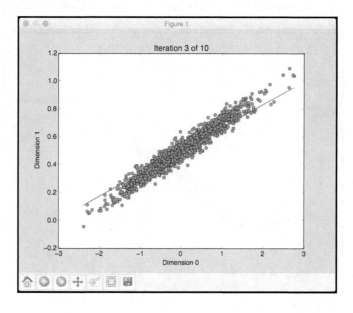

It looks like the line is getting closer to the real model. If you continue iterating like this, the model gets better. The eighth iteration looks like this:

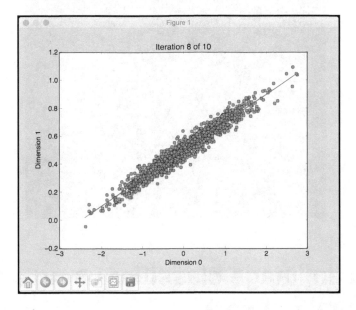

The line seems to fit the data pretty well. You will see the following printed on your Terminal in the beginning:

```
ITERATION 1
W = -0.130961
b = 0.53005
loss = 0.0760343

ITERATION 2
W = 0.0917911
b = 0.508959
loss = 0.00960302

ITERATION 3
W = 0.164665
b = 0.502555
loss = 0.00250165

ITERATION 4
W = 0.188492
b = 0.500459
loss = 0.0017425
```

Once it finishes training, you will see the following on your Terminal:

```
ITERATION 7
W = 0.199662
b = 0.499477
loss = 0.00165175

ITERATION 8
W = 0.199934
b = 0.499453
loss = 0.00165165

ITERATION 9
W = 0.200023
b = 0.499445
loss = 0.00165164

ITERATION 10
W = 0.200052
b = 0.499443
loss = 0.00165164
```

Building an image classifier using a single layer neural network

Let's see how to create a single layer neural network using `TensorFlow` and use it to build an image classifier. We will be using MNIST image dataset to build our system. It is dataset containing handwritten images of digits. Our goal is to build a classifier that can correctly identify the digit in each image.

Create a new python and import the following packages:

```
import argparse

import tensorflow as tf
from tensorflow.examples.tutorials.mnist import input_data
```

Define a function to parse the input arguments:

```
def build_arg_parser():
    parser = argparse.ArgumentParser(description='Build a classifier using \
            MNIST data')
    parser.add_argument('--input-dir', dest='input_dir', type=str,
            default='./mnist_data', help='Directory for storing data')
    return parser
```

Define the main function and parse the input arguments:

```
if __name__ == '__main__':
    args = build_arg_parser().parse_args()
```

Extract the MNIST image data. The one_hot flag specifies that we will be using one-hot encoding in our labels. It means that if we have n classes, then the label for a given datapoint will be an array of length n. Each element in this array corresponds to a particular class. To specify a class, the value at the corresponding index will be set to 1 and everything else will be 0:

```
# Get the MNIST data
mnist = input_data.read_data_sets(args.input_dir, one_hot=True)
```

The images in the database are *28 x 28*. We need to convert it to a single dimensional array to create the input layer:

```
# The images are 28x28, so create the input layer
# with 784 neurons (28x28=784)
x = tf.placeholder(tf.float32, [None, 784])
```

Create a single layer neural network with weights and biases. There are 10 distinct digits in the database. The number of neurons in the input layer is *784* and the number of neurons in the output layer is *10*:

```
# Create a layer with weights and biases. There are 10 distinct
# digits, so the output layer should have 10 classes
W = tf.Variable(tf.zeros([784, 10]))
b = tf.Variable(tf.zeros([10]))
```

Create the equation to be used for training:

```
# Create the equation for 'y' using y = W*x + b
y = tf.matmul(x, W) + b
```

Define the loss function and the gradient descent optimizer:

```
# Define the entropy loss and the gradient descent optimizer
y_loss = tf.placeholder(tf.float32, [None, 10])
loss = tf.reduce_mean(tf.nn.softmax_cross_entropy_with_logits(y, y_loss))
optimizer = tf.train.GradientDescentOptimizer(0.5).minimize(loss)
```

Initialize all the variables:

```
# Initialize all the variables
init = tf.initialize_all_variables()
```

Create a `TensorFlow` session and run it:

```
# Create a session
session = tf.Session()
session.run(init)
```

Start the training process. We will train using batches where we run the optimizer on the current batch and then continue with the next batch for the next iteration. The first step in each iteration is to get the next batch of images to train on:

```
# Start training
num_iterations = 1200
batch_size = 90
for _ in range(num_iterations):
    # Get the next batch of images
    x_batch, y_batch = mnist.train.next_batch(batch_size)
```

Run the optimizer on this batch of images:

```
# Train on this batch of images
session.run(optimizer, feed_dict = {x: x_batch, y_loss: y_batch})
```

Once the training process is over, compute the accuracy using the test dataset:

```
# Compute the accuracy using test data
predicted = tf.equal(tf.argmax(y, 1), tf.argmax(y_loss, 1))
accuracy = tf.reduce_mean(tf.cast(predicted, tf.float32))
print('\nAccuracy =', session.run(accuracy, feed_dict = {
        x: mnist.test.images,
        y_loss: mnist.test.labels}))
```

The full code is given in the file `single_layer.py`. If you run the code, it will download the data to a folder called `mnist_data` in the current folder. This is the default option. If you want to change it, you can do so using the input argument. Once you run the code, you will get the following output on your Terminal:

```
Extracting ./mnist_data/train-images-idx3-ubyte.gz
Extracting ./mnist_data/train-labels-idx1-ubyte.gz
Extracting ./mnist_data/t10k-images-idx3-ubyte.gz
Extracting ./mnist_data/t10k-labels-idx1-ubyte.gz

Accuracy = 0.921
```

As printed on your Terminal, the accuracy of the model is 92.1%.

Building an image classifier using a Convolutional Neural Network

The image classifier in the previous section didn't perform well. Getting *92.1%* on MNIST dataset is relatively easy. Let's see how we can use Convolutional Neural Networks (CNNs) to achieve a much higher accuracy. We will build an image classifier using the same dataset, but with a CNN instead of a single layer neural network.

Create a new python and import the following packages:

```
import argparse

import tensorflow as tf
from tensorflow.examples.tutorials.mnist import input_data
```

Define a function to parse the input arguments:

```
def build_arg_parser():
    parser = argparse.ArgumentParser(description='Build a CNN classifier \
            using MNIST data')
    parser.add_argument('--input-dir', dest='input_dir', type=str,
            default='./mnist_data', help='Directory for storing data')
    return parser
```

Define a function to create values for weights in each layer:

```
def get_weights(shape):
    data = tf.truncated_normal(shape, stddev=0.1)
    return tf.Variable(data)
```

Define a function to create values for biases in each layer:

```
def get_biases(shape):
    data = tf.constant(0.1, shape=shape)
    return tf.Variable(data)
```

Define a function to create a layer based on the input shape:

```
def create_layer(shape):
    # Get the weights and biases
    W = get_weights(shape)
    b = get_biases([shape[-1]])

    return W, b
```

Define a function to perform 2D-convolution:

```
def convolution_2d(x, W):
    return tf.nn.conv2d(x, W, strides=[1, 1, 1, 1],
            padding='SAME')
```

Define a function to perform a 2×2 max pooling operation:

```
def max_pooling(x):
    return tf.nn.max_pool(x, ksize=[1, 2, 2, 1],
            strides=[1, 2, 2, 1], padding='SAME')
```

Define the main function and parse the input arguments:

```
if __name__ == '__main__':
    args = build_arg_parser().parse_args()
```

Extract the *MNIST* image data:

```
# Get the MNIST data
mnist = input_data.read_data_sets(args.input_dir, one_hot=True)
```

Create the input layer with *784* neurons:

```
# The images are 28x28, so create the input layer
# with 784 neurons (28x28=784)
x = tf.placeholder(tf.float32, [None, 784])
```

We will be using convolutional neural networks that take advantage of the 2D structure of images. So let's reshape x into a 4D tensor where the second and third dimensions specify the image dimensions:

```
# Reshape 'x' into a 4D tensor
x_image = tf.reshape(x, [-1, 28, 28, 1])
```

Create the first convolutional layer that will extract *32* features for each *5×5* patch in the image:

```
# Define the first convolutional layer
W_conv1, b_conv1 = create_layer([5, 5, 1, 32])
```

Convolve the image with weight tensor computed in the previous step, and then add the bias tensor to it. We then need to apply the **Rectified Linear Unit (ReLU)** function to the output:

```
# Convolve the image with weight tensor, add the
# bias, and then apply the ReLU function
h_conv1 = tf.nn.relu(convolution_2d(x_image, W_conv1) + b_conv1)
```

Apply the 2×2 max pooling operator to the output of the previous step:

```
# Apply the max pooling operator
h_pool1 = max_pooling(h_conv1)
```

Create the second convolutional layer to compute 64 features for each 5×5 patch:

```
# Define the second convolutional layer
W_conv2, b_conv2 = create_layer([5, 5, 32, 64])
```

Convolve the output of the previous layer with weight tensor computed in the previous step, and then add the bias tensor to it. We then need to apply the **Rectified Linear Unit (ReLU)** function to the output:

```
# Convolve the output of previous layer with the
# weight tensor, add the bias, and then apply
# the ReLU function
h_conv2 = tf.nn.relu(convolution_2d(h_pool1, W_conv2) + b_conv2)
```

Apply the 2×2 max pooling operator to the output of the previous step:

```
# Apply the max pooling operator
h_pool2 = max_pooling(h_conv2)
```

The image size is now reduced to 7×7. Create a fully connected layer with 1024 neurons.

```
# Define the fully connected layer
W_fc1, b_fc1 = create_layer([7 * 7 * 64, 1024])
```

Reshape the output of the previous layer:

```
# Reshape the output of the previous layer
h_pool2_flat = tf.reshape(h_pool2, [-1, 7*7*64])
```

Multiply the output of the previous layer with the weight tensor of the fully connected layer, and then add the bias tensor to it. We then apply the Rectified Linear Unit (ReLU) function to the output:

```
# Multiply the output of previous layer by the
# weight tensor, add the bias, and then apply
# the ReLU function
h_fc1 = tf.nn.relu(tf.matmul(h_pool2_flat, W_fc1) + b_fc1)
```

In order to reduce overfitting, we need to create a dropout layer. Let's create a `TensorFlow` placeholder for the probability values that specify the probability of a neuron's output being kept during dropout:

```
# Define the dropout layer using a probability placeholder
# for all the neurons
keep_prob = tf.placeholder(tf.float32)
h_fc1_drop = tf.nn.dropout(h_fc1, keep_prob)
```

Define the readout layer with 10 output neurons corresponding to 10 classes in our dataset. Compute the output:

```
# Define the readout layer (output layer)
W_fc2, b_fc2 = create_layer([1024, 10])
y_conv = tf.matmul(h_fc1_drop, W_fc2) + b_fc2
```

Define the loss function and optimizer function:

```
# Define the entropy loss and the optimizer
y_loss = tf.placeholder(tf.float32, [None, 10])
loss = tf.reduce_mean(tf.nn.softmax_cross_entropy_with_logits(y_conv, y_loss))
optimizer = tf.train.AdamOptimizer(1e-4).minimize(loss)
```

Define how the accuracy should be computed:

```
# Define the accuracy computation
predicted = tf.equal(tf.argmax(y_conv, 1), tf.argmax(y_loss, 1))
accuracy = tf.reduce_mean(tf.cast(predicted, tf.float32))
```

Create and run a session after initializing the variables:

```
# Create and run a session
sess = tf.InteractiveSession()
init = tf.initialize_all_variables()
sess.run(init)
```

Start the training process:

```
# Start training
num_iterations = 21000
batch_size = 75
print('\nTraining the model....')
for i in range(num_iterations):
    # Get the next batch of images
    batch = mnist.train.next_batch(batch_size)
```

Print the accuracy progress every 50 iterations:

```
        # Print progress
        if i % 50 == 0:
            cur_accuracy = accuracy.eval(feed_dict = {
                    x: batch[0], y_loss: batch[1], keep_prob: 1.0})
            print('Iteration', i, ', Accuracy =', cur_accuracy)
```

Run the optimizer on the current batch:

```
        # Train on the current batch
        optimizer.run(feed_dict = {x: batch[0], y_loss: batch[1],
    keep_prob: 0.5})
```

Once the training process is over, compute the accuracy using the test dataset:

```
    # Compute accuracy using test data
    print('Test accuracy =', accuracy.eval(feed_dict = {
            x: mnist.test.images, y_loss: mnist.test.labels,
            keep_prob: 1.0}))
```

The full code is given in the file cnn.py. If you run the code, you will get the following output on your Terminal:

```
Extracting ./mnist_data/train-images-idx3-ubyte.gz
Extracting ./mnist_data/train-labels-idx1-ubyte.gz
Extracting ./mnist_data/t10k-images-idx3-ubyte.gz
Extracting ./mnist_data/t10k-labels-idx1-ubyte.gz

Training the model....
Iteration 0 , Accuracy = 0.0533333
Iteration 50 , Accuracy = 0.813333
Iteration 100 , Accuracy = 0.8
Iteration 150 , Accuracy = 0.906667
Iteration 200 , Accuracy = 0.84
Iteration 250 , Accuracy = 0.92
Iteration 300 , Accuracy = 0.933333
Iteration 350 , Accuracy = 0.866667
Iteration 400 , Accuracy = 0.973333
Iteration 450 , Accuracy = 0.933333
Iteration 500 , Accuracy = 0.906667
Iteration 550 , Accuracy = 0.853333
Iteration 600 , Accuracy = 0.973333
Iteration 650 , Accuracy = 0.973333
Iteration 700 , Accuracy = 0.96
Iteration 750 , Accuracy = 0.933333
```

As you continue iterating, the accuracy keeps increasing as shown in the following screenshot:

```
Iteration 2900 , Accuracy = 0.973333
Iteration 2950 , Accuracy = 1.0
Iteration 3000 , Accuracy = 0.973333
Iteration 3050 , Accuracy = 1.0
Iteration 3100 , Accuracy = 0.986667
Iteration 3150 , Accuracy = 1.0
Iteration 3200 , Accuracy = 1.0
Iteration 3250 , Accuracy = 1.0
Iteration 3300 , Accuracy = 1.0
Iteration 3350 , Accuracy = 1.0
Iteration 3400 , Accuracy = 0.986667
Iteration 3450 , Accuracy = 0.946667
Iteration 3500 , Accuracy = 0.973333
Iteration 3550 , Accuracy = 0.973333
Iteration 3600 , Accuracy = 1.0
Iteration 3650 , Accuracy = 0.986667
Iteration 3700 , Accuracy = 1.0
Iteration 3750 , Accuracy = 1.0
Iteration 3800 , Accuracy = 0.986667
Iteration 3850 , Accuracy = 0.986667
Iteration 3900 , Accuracy = 1.0
```

Now that we have the output, we can see that the accuracy of a convolutional neural network is much higher than a simple neural network.

Summary

In this chapter, we learnt about Deep Learning and CNNs. We discussed what CNNs are and why we need them. We talked about the architecture of CNNs. We learnt about the various type of layers used within a CNN. We discussed how to use TensorFlow. We used it to build a perceptron-based linear regressor. We learnt how to build an image classifier using a single layer neural network. We then built an image classifier using a CNN.

Index

1
15-puzzle
  URL 189

8
8-puzzle solver
  building 189, 191

A
A* algorithm 190
Affinity Propagation model
  about 120
  used, for obtaining subgroups in stock market 120, 121, 122
Alpha-Beta pruning 235, 236
alphabet sequences
  identifying, with Conditional Random Fields (CRFs) 296, 299
AlphaGo 13
analytical models 24
annealing schedule 178
Artificial Intelligence (AI), applications
  Computer Vision 12
  Expert Systems 13
  Games 13
  Natural Language Processing 13
  Robotics 14
  Speech Recognition 13
Artificial Intelligence (AI), branches
  genetic programming 16
  heuristic 16
  knowledge representation 15
  logic-based AI 15
  machine learning 14
  pattern recognition 14
  planning 16
  search techniques 15
Artificial Intelligence (AI)
  about 7, 8
  need for 8, 11
artificial neural networks
  about 363, 364
  building 364
  training 364
audio file dataset
  URL 320
audio signals
  generating 311
  transforming, to frequency domain 309, 311
  visualizing 306, 307, 308

B
background subtraction
  used, for object tracking 335
Bag of Words model
  used, for extracting frequency of terms 262, 263, 265
binarization 33
binary classification, output
  false negatives 47
  false positives 47
  true negatives 47
  true positives 47
bit pattern
  generating, with predefined parameters 203
bot
  building, for Last Coin Standing 237
  building, for Tic-Tac-Toe 241, 243
Breadth First Search (BFS) 176

C
CAMShift algorithm
  about 339

reference link 340
used, for building object tracker 339
category predictor
about 265
building 265, 266, 267, 268
census income dataset
URL 52
characters
visualizing, in OCR database 386
class imbalance
dealing with 82, 83, 85, 86, 87, 89
classification 32
Cognitive Modeling 19
collaborative filtering
about 145
used, for obtaining similar users 145, 146, 148
colorspaces
used, for tracking objects 331, 332, 333, 334, 335
combinatorial search 234, 235
Conditional Random Fields (CRFs)
about 296
used, for identifying alphabet sequences 296, 299
confusion matrix 47, 48, 50
Constraint Satisfaction Problems (CSPs) 177
constraints
problem, solving 183
Convolutional Neural Networks (CNNs)
about 407, 408
architecture 408
convolutional layer 409
fully connected layer 409
input layer 409
pooling layer 409
rectified linear unit layer 409
used, for building image classifier 419, 420
Covariance Matrix Adaptation Evolution Strategy (CMA-ES) 210
cvxopt
URL 280

D

data clustering
with K-Means algorithm 100, 101, 102, 104, 105
data preprocessing
about 33
binarization 33
mean, removing 34
normalization 36
scaling 35
data
generating, with Hidden Markov Model (HMM) 292, 294
loading 27, 28, 29
DEAP package
URL 203
decision tree
about 66
classifier, building 67, 68, 70, 71
URL 67
Depth First Search (DFS) 176
Dijkstra's algorithm 190
discrete cosine transform (DCT) 317

E

easyAI library
installing 236, 237
URL 236, 237
eigenvalues
reference link 117
eigenvectors
reference link 117
ensemble learning
about 65
learning models, building 66
Euclidean distance
URL 141
Euclidean score 141
evolution
visualizing 210, 216, 217, 218
evolutionary algorithm 201, 202
Expectation-Maximization (EM) 115
Extremely Random Forests
about 72
classifier, building 72, 73, 75, 76, 78
regressor, building for traffic prediction 95, 96, 98
eye detection 358

F

face detection
  about 354
  Haar cascade, used for object detection 354
  integral images, using for feature extraction 355
family tree
  parsing 161, 165
fitness function 202
Fourier Transform
  about 309
  URL 309
frame differencing 328
frequency domain
  audio signals, transforming 309, 311

G

Gaussian Mixture Models (GMMs)
  about 114
  classifier, building 115, 117, 119
gender identifier
  constructing 268, 269, 270, 271
General Problem Solver (GPS)
  about 21
  used, for solving problem 22
genetic algorithm
  about 201, 202
  concepts 202, 203
geography
  analyzing 167, 168
greedy search
  about 179
  used, for constructing string 179, 180
grid search
  used, for obtaining optimal training parameters 89, 90, 91, 92

H

Haar cascades
  used, for object detection 354
heuristic 176
heuristic search
  about 175, 176
  uninformed, versus informed search 176
Hexapawn

  about 248
  multiple bots, building for 248
Hidden Markov Model (HMM)
  about 292
  reference link 320
  used, for generating data 292
hmmlearn package
  URL 320
housing prices
  estimating, with Support Vector Regressor 61, 62
hyperplane 50

I

image classifier
  building, with Convolutional Neural Networks (CNNs) 419, 420
  building, with single layer neural network 416, 417, 418
image convolution
  URL 409
income data
  classifying, with Support Vector Machine (SVM) 52, 53, 55
Information Processing Language (IPL) 21
informed search
  about 176
  versus uninformed search 176
integral images
  using, for feature extraction 355
intelligence
  defining, with Turing Test 16
intelligent agent
  building 22, 23
  models 24
interactive object tracker
  building, with CAMShift algorithm 339, 340

K

K-Means algorithm
  used, for data clustering 100, 101, 102, 104, 105
K-Nearest Neighbors classifier
  building 134, 135, 136, 140

L

L1 normalization 36
L2 normalization 36
label encoding 37, 38
Last Coin Standing
  bot, building for 237
Latent Dirichlet Allocation
  about 275
  used, for topic modeling 275, 278
learned models 24
learning agent
  building 402, 404
learning models
  building, with ensemble learning 66
least absolute deviations 36
lemmatization
  about 258
  used, for converting word to base forms 258, 259
local search techniques
  about 177
  simulated annealing 178, 179
logic programming
  about 153, 154, 155
  building blocks 156
  used, for problem solving 156, 157
logistic regression classifier 38, 39, 41, 42
logpy package
  URL 157
Lucas-Kanade method
  URL 347

M

Mac OS X
  Python 3, installing 25
machines
  with human thinking capability 18, 19
mathematical expressions
  matching 157
matplotlib
  URL 26
matrix operations
  reference link 293
Maximum A-Posteriori (MAP) 115
maximum margin 51
maze solver
  building 194
mean removal 34
Mean Shift algorithm
  about 106
  used, for estimating number of clusters 106, 107, 108, 109
  used, for segmenting market based on shopping patterns 122, 124, 126
mean squared error (MSE) 220
Mel Frequency Cepstral Coefficients (MFCCs)
  about 316
  reference link 317
Minimax algorithm 235
models
  analytical models 24
  learned models 24
movie recommendation system
  building 148, 149, 150, 151
multilayer neural network
  constructing 373
multiple bots
  building, for Hexapawn 248
multivariable regressor
  building 59, 60, 61
music
  generating, via synthesizing tones 314, 316

N

Natural Language Processing (NLP)
  about 253
  packages, installing 253
Natural Language Toolkit (NLTK)
  about 254
  URL 254
Naïve Bayes classifier 43, 45, 47
nearest neighbors
  extracting 130, 131, 133, 134
Negamax algorithm 236
neural networks
  reference link 364
NeuroLab
  URL 365
normalization 36

[428]

NumPy
  URL 26

O

object detection
  Haar cascades, using 354
  URL 354
objects
  tracking, with background subtraction 335
  tracking, with colorspaces 331, 332, 333, 334, 335
OCR database
  characters, visualizing 386
  URL 386
One Max problem 204
OpenAI Gym package
  URL 397
OpenCV 3
  URL, for installation on various OS 328
OpenCV
  about 328
  installing 328
  URL 328
Optical Character Recognition (OCR)
  about 386
  engine, building 388
optical flow
  used, for tracking 347

P

packages
  installing 26
Pandas
  time-series data, handling 280, 283
Pearson score 141
Perceptron 365
Perceptron based classifier
  building 365, 366
perceptron-based linear regressor
  building 410, 412, 415
prime numbers
  validating 159, 160
programming paradigms
  declarative 154
  functional 154
  imperative 154
  logic 154
  object oriented 154
  procedural 154
  symbolic 154
pruning 236
puzzle solver
  building 170
Python 3
  installing 24
  installing, on Mac OS X 25
  installing, on Ubuntu 25
  installing, on Windows 26
  URL, for installation 26
Python
  packages, installing 157
python_speech_features package
  URL 317

R

Random Forest
  about 72
  classifier, building 72, 73, 75, 76, 78
  confidence measure, estimating of predictions 78, 80, 82
rational agents
  building 20, 21
Rectified Linear Unit (ReLU) 420
recurrent neural networks
  URL 381
  used, for analyzing sequential data 381
region-coloring problem
  solving 186
regression 55
reinforcement learning, examples
  babies 395
  game playing 395
  industrial controllers 395
  robotics 395
reinforcement learning
  about 394
  building blocks 396
  environment, creating 397, 399, 400, 401
  examples 395
  versus supervised learning 394

relative feature
  importance, computing 92, 94, 95
respondent machine, Turing Test
  Knowledge Representation 17
  Machine Learning 18
  Natural Language Processing 17
  Reasoning 18
robot controller
  building 224, 225, 231

S

scaling 35
scikit-learn
  URL 26
SciPy-stack compatible distribution
  URL 26
SciPy
  URL 26
search algorithms
  using, in games 234
sentiment analyzer
  building 271, 272, 273, 274
sequential data
  about 279, 280
  analyzing, with recurrent neural networks 381, 383
sigmoid curve 38
silhouette scores
  used, for estimating quality of clustering 109, 110, 112, 113, 114
similarity scores
  computing 141, 142, 145
simple AI
  URL 179
Simulated Annealing 178, 179
single layer neural network
  constructing 369
  used, for building image classifier 416, 417, 418
single variable regressor
  building 56, 57, 59
sinusoids 311
speech features
  extracting 316, 317, 318, 319
speech recognition 305
speech signals 305, 306

spoken words
  recognizing 320, 321, 322, 326
stemming
  used, for converting words to base forms 256, 257, 258
stock market data
  analyzing 300, 301, 302
supervised learning
  about 32
  versus reinforcement learning 394
  versus unsupervised learning 32
Support Vector Machine (SVM)
  about 50, 51
  used, for classifying income data 52, 53, 55
Support Vector Regressor
  housing prices, estimating 61, 62
Support Vectors 51
survival of the fittest approach 202
symbol regression problem
  solving 219

T

TensorFlow
  URL 410
term frequency
  extracting, with Bag of Words model 262, 263, 265
TermFrequency - Inverse Document Frequency (tf-idf) 265
text data
  dividing, into chunks 260
  tokenizing 255, 256
Tic-Tac-Toe
  about 234
  bot, building for 241, 243
time-series data
  handling, with Pandas 280, 283
  operating on 285, 286, 287
  slicing 283
  statistics, extracting 288, 289, 290, 291
Tkinter package
  URL 38
tokenization 255
tones
  reference link 314

synthesizing, for music generation 314, 315, 316
topic modeling
  about 275
  with Latent Dirichlet Allocation 275, 277, 278
traffic dataset
  URL 95
training pipeline
  creating 127, 130
Turing Test
  used, for defining intelligence 16, 18

U

Ubuntu
  Python 3, installing 25
Uniform Cost Search (UCS) 176
uninformed search
  versus informed search 176

unsupervised learning
  about 32, 99
  versus supervised learning 32

V

Vector Quantization 378
vector quantizer
  building 378

W

Windows
  Python 3, installing 26
words
  converting, to base forms with lemmatization 258, 259
  converting, to base forms with stemming 256, 257, 258